IET MATERIALS, CIRCUITS AND DEVICES SERIES 75

Understandable Electronic Devices

Other volumes in this series:

Understandable Electronic Devices

Key concepts and
circuit design

Meizhong Wang

The Institution of Engineering and Technology

Published by The Institution of Engineering and Technology, London, United Kingdom

The Institution of Engineering and Technology is registered as a Charity in England & Wales (no. 211014) and Scotland (no. SC038698).

© The Institution of Engineering and Technology 2022

First published 2022

The Institution of Engineering and Technology
Futures Place Six Hills Way,
Stevenage Herts, SG1 2UA,
United Kingdom

www.theiet.org

British Library Cataloguing in Publication Data
A catalogue record for this product is available from the British Library

ISBN 978-1-83953-216-0 (hardback)
ISBN 978-1-83953-217-7 (PDF)

Typeset in India by Exeter Premedia Services

Cover Image: Richard Newstead via Getty Images

Table of contents

Preface

If you are looking for a quick, concise, homework/exam guide, and review book in Electronic Circuits and Devices, this unique and well-structured *Understandable Electronic Circuits and Devices – Key Concepts* is an excellent supplement and convenient reference resource. Skip the lengthy and distracting books and instead use this concise book as a guideline for your studies, quick reviewing, or tutoring. This book provides a quick, understandable, and effective guide on electronic circuits and devices.

Key features
As an aid to readers, the book provides some noteworthy features:
- A concise study guide, quickly getting to the heart of each topic, helping readers with a quick review (or for students before doing homework as well as preparation for exams).
- Clear and easy-to-understand written format and style. Materials are presented in visual and grayscale format with less text and more outlines, tables, etc.; clearly presenting information and making studying/reviewing more effective.
- Key terms, properties, phrases, concepts, formulas, etc. are easily located. Clear step-by-step procedures for applying theorems.
- Summary at the end of each chapter to emphasize the key points, formulas, figures, etc. in the chapter.
- Self-tests after each chapter will help readers focus on the key principles, complete the connection between theory and practice, and assist readers in the learning process.

Suitable readers
This book is intended for college/university students, technicians, technologists, engineers, or any other professionals who require a solid foundation in the basics of electronic circuits and devices.

It targets an audience of all sectors in the fields of electronics and computer engineering such as electronics, computer, communications, control and automation, embedded systems, signal processing, power electronics, industrial instrumentation, power systems (including renewable energy), electrical apparatus and machines, nanotechnology, biomedical imaging, information technology, artificial intelligence and more. It is also suitable for non-electronics readers. It provides readers with the necessary foundation for electronic circuits and devices in related fields.

To make this book more reader-friendly, the concepts, new terms, laws/rules, and theorems are explained in an easy-to-understand style. Clear step-by-step procedures for applying methods of electronics theorems make this book easy for readers to learn electronic circuits and devices themselves.

Acknowledgments

Special thanks to Sarah Lynch, commissioning editor at the Institution of Engineering and Technology (IET). I really appreciate her belief in my ability to write this book and her help and support in publishing it. I also appreciate the support from Olivia Wilkins, assistant editor, Bianca Campbell, books and journals sales manager, Suzanne Bishop, marketing manager, Felicity Hull, marketing executive, Jo Hughes, production controller, and Paul Deards, publisher at IET.

Reference

I used *Electronic Devices – Conventional Current Version authored by Thomas L. Floyd* as text when I taught an Electronics course at a college in Canada. This *Understandable Electronic Circuits and Devices – Key concepts* is based on my teaching notes from this course and text.

About the author

Meizhong Wang is a recently retired college instructor, and she has been an instructor at the College of New Caledonia (CNC) in Canada for 31 years. She has lectured in electronics, electric circuits, math, physics, computers, , etc. at CNC and other colleges and universities in Canada and China.

https://bccampus.ca/2021/05/25/bccampus-award-for-excellence-in-open-education-mei-wang/
https://cnc.bc.ca/news/detail/2021/08/04/cnc-instructor-receives-bccampus-award

Meizhong is also the author of several books, including:

- *Middle School Science – Key Concepts, Practice, and Quizzes* (The Critical Thinking Co. – U.S., Sept. 2021).
- *Key Concepts of Computer* (BCcampus – Canada, Dec. 2020)
- *Key Concepts of Intermediate Level Math* (BCcampus – Canada, 2018)
- *Algebra I & II – Key Concepts, Practice, and Quizzes* (The Critical Thinking Co. – U.S., 2013).
- *Math Made Easy* (CNC Press, Canada, 2011, second edition 2013).
- *Understandable Electric Circuits* (Michael Faraday House of the IET – Institution of Engineering and Technology – U.K., 2010, second edition 2018).
- *Legends of Four Chinese Sages* – Coauthor (Lily S.S.C Literary Ltd. – Canada, 2007).
- 简明电路基础, Chinese version of Understandable Electric Circuits (The Higher Education Press – China, 2005).

Introduction

Why Study Electronic Circuits?

- Electronics has enormously shaped the growth of modern society. It is the great driving force for modern industry and civilization. It dominates the world today and the people in it.
- Our everyday life would be nearly unthinkable without electronics. Electronic devices and gadgets (such as mobile phone, computer, internet, television, home office equipment, home entertainment equipment, home appliances, ...) permeate our daily life to a large extent.
- Electronic professionals work at the forefront of electronic technologies, designing and improving the electronic devices and gadgets we use daily. They innovate to meet society's communication, technology, energy, daily life, etc. needs.
- Electronic engineering has made and continues to make incredible contributions to most aspects of human society – a truth that cannot be neglected. Moreover, it may have a bigger impact on human civilization in the future.
- Since the increase in interest and the rise of computer technology, artificial intelligence, quantum computing, robotics, smart and autonomous systems, etc., electronic circuits and devices are playing an important role in the digital age.
- When you start reading this book, perhaps you have already chosen the electronic fields as your professional goal – a wise choice! There are lots of career opportunities in this field.
- Experts forecast that the demand for professionals in this field will have a huge demand. This is good news for people who have chosen these areas of study and you will join people at the forefront of technologies and innovations the elite of the world.
- Reading this book or other electronic books is the first step into the electronic world that will introduce you to the foundation of the professions in these areas.

Applications of electronics

- Consumer electronics
  - Office gadgets: computer, internet, home office equipment (scanner/printer, FAX machine), etc.
  - Home electronics: outdoor security camera, smart home, robot vacuum cleaner, smart faucet, home appliances, home entertainment equipment, etc.
  - Audio and video systems: mobile phone, tablet, headphone, TV, microphone/loudspeaker, video game console, digital camera, etc.
  - ... ...

Communication devices
- Infrared devices (remote control, infrared therapy, infrared thermometer, etc.)
- Network card
- Wi-Fi devices
- Fiber-optic communication devices
- Bluetooth devices

It connects short-range devices to communicate with each other without wires.
- Intelligent virtual assistant (IVA)
- It can perform tasks or services for assisting humans in a variety of tasks like customer service (such as IT helpdesk, HR, finance, weather, sports, movies, etc.)
- Popular examples: Google Assistant, Cortana, Siri, Alexa, Nina, etc.
- ... ...

Industrial electronics
- Automation and motion control
- Image processing
- Mechatronics and robotics
- Renewable energy applications
- Photo voltaic systems
- Biomechanics
- Motor drive control
- ... ...

Automotive
- Engine control unit
- Transmission electronics
- Passive safety
- Driver assistance
- Entertainment systems
- Intelligent vehicle technologies
- Satellite navigation
- ... ...

The vehicle's computer system is made up of electronic control units that are accompanied by many different sensors to control/monitor different areas of the vehicles (engine, transmission, brake system, communication, safety, etc.).

Military electronics
- Missile launching system
- Aircraft system
- Cockpit controller
- Military radar
- Rocket launchers
- ... ...

Medical electronic devices

- Magnetic resonance imaging (MRI)
- Blood gas analyzer
- Glucose meter (or monitor)
- Heart monitor
- Infrared and digital thermometer
- Respiration monitor
- Electronic brain wave machine
  A device that measures electrical activity in the brain.
- Pacemaker
  A small electronic device that's implanted in the chest to help regulate the heartbeat.
- Defibrillator
  It causes electrical shock to heart muscles and brings back the heart to normal working condition.
- ... ...

Aerospace

- Integrated flight system
- Communication device
- Adequate navigation
- Power supply
- Thermal considerations
- Fuel conservation
- Safety against vibration and shock
- ... ...

Careers in Electronic Engineering

- Nowadays, electronics engineering technology is developing so rapidly that many career options exist for those who have chosen this field.
- As long as you have gained a solid foundation in electronics, the training that most employers provide in their branches will lead you into a brand new professional career very quickly.
- There are many types of jobs for electronic engineering technology. Only a partial list is as follows:
  - Electronics engineer
  - Control and automation engineer
  - Process and system engineer
  - Instrument engineer
  - Robotics engineer
  - Product engineer
  - Field engineer
  - Reliability engineer
  - Integrated circuits (IC) design engineer

- Computer engineer
- Power electronics engineer
- Electronic engineering professor/lecturer
- System engineer
- Electronic test engineer/technician
- Solar photovoltaic technician
- Designer and technologist
- Electronics technician
- Equipment maintenance technician
- Calibration/lab technician
- Technical writer for electronic products
- Electronics repair
- ... ...

- Electronics technicians, technologists, engineers, and experts will be in demand in the future, so you definitely do not want to miss this good opportunity.

Chapter 1

Basic semiconductor theory

Chapter outline

1.1   Fundamental semiconductor theory

1.1.1   Basic atomic theory

Atomic structure

- Atom: the smallest particle of matter (or substance) that has the chemical properties of that matter.

  It means that an atom of water is still water, an atom of oxygen is still oxygen, etc.

- Atoms are made up of three parts:
  - Proton (p^+): a positively charged particle (it has a mass = 1 amu).

    Amu: the unit of measurement for atoms or particles.
  - Electron (e^-): a negatively charged particle (it has a mass $\approx$ 0).
  - Neutron (n): an uncharged (neutral) particle (it has a mass = 1 amu).
- A neutral atom has an equal number of electrons and protons.

  No. of p^+ = No. of e^-
- Nucleus (or core): the center core of an atom that contains protons and neutrons.
- Electrons in the shells that orbit the nucleus of an atom like planets orbit the sun.

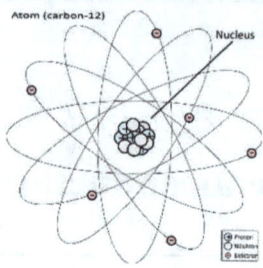

Figure 1.1   Atoms

Table 1.1   Particle

Particle name	Symbol	Charge	Mass	Location
Proton	p^+	Positive +	1 amu	Nucleus
Electron	e^-	Negative −	0	Orbitals
Neutron	n	Neutral	1 amu	Nucleus

Ion

- Ion: an atom that has gained or lost electron(s). It is an atom in which the number of electrons is different from the number of protons.
  - In a standard neutral atom: No. of electrons = No. of protons
  - In an ion: No. of electrons ≠ No. of protons
- Positive ion: a neutral atom *loses* an electron.
- Negative ion: a neutral atom *gains* an electron.

Electrons in an atom exist in shells (or energy levels)

- The closer the electron is to the nucleus, the higher the binding energy it has.
- The farther the electron is to the nucleus, the lower the binding energy it has (the easier it is to take away from the atom).

Bohr diagram (Bohr model)

- The nucleus is at the center and the electrons are located on different shells.
- Valence electrons (or free electrons): the electrons in the outmost shell (or valence shell) of an atom.

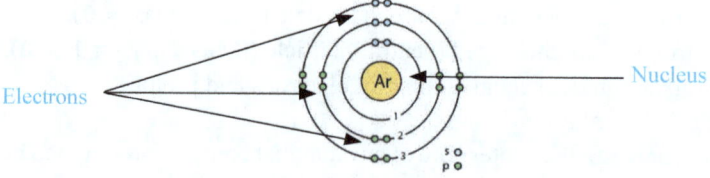

Figure 1.2   Bohr diagram – an example

Arrangements of electrons

• Electrons are located in shells according to rules:
  – The first shell can only hold two electrons.
  – The second shell can hold up to (maximum capacity) eight electrons.
  – The third shell can hold up to 18 electrons.
  – The fourth shell can hold up to 32 electrons.
  – ......
  – The maximum number of electrons in the outmost shell (valence shell) is 8.

    Atoms fill up their electron shells from the inside out (1st, 2nd,…).

Example: The element gold (Au):

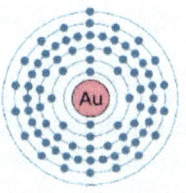

Figure 1.3   The element gold

• Octet rule: the tendency of atoms to have eight electrons in the valence shell (outmost shell).

  Every atom wants to have eight valence electrons in its outermost shell to be most stable.

• Atoms that do not have eight electrons in their outer shell will tend to gain or lose electrons.
• Atoms with more than four valence electrons tend to gain electrons.
• Atoms with less than four valence electrons tend to lose electrons.

Example:

– The element oxygen (O): atomic no. = 8

    valence electrons = 6

Oxygen tends to gain two electrons in the valence shell.

– The element sodium (Na): atomic no. = 11

    valence electrons = 1

Sodium tends to lose one electron.

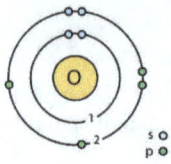

Figure1.4   Oxygen

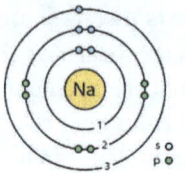

Figure 1.5   Sodium

1.1.2   *N-type and P-type semiconductors*

Semiconductor

- Conductor: a material that allows electric current or heat to flow easily (most metals are good conductors).
- Insulator: a material that prevents the flow of electric current or heat (most non-metals are good insulators).
- Semiconductor: a material with electrical conductivity between that of an insulator and a conductor (it is neither a conductor nor an insulator). It is a material that allows some current to flow and is used especially in electronic devices.
- The elements silicon and germanium are semiconductors.
- The elements silicon (Si) and germanium (Ge) have four valence electrons which can be shared with neighboring atoms.

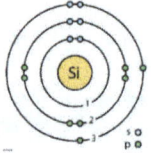

Figure 1.6   Silicon

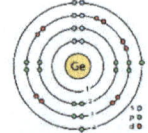

Figure 1.7   Germanium

Electron-hole pair

- Hole: a hole corresponds to the vacancy created by a free electron in a semiconductor.
  - When an electron breaks away to become free, it leaves a hole in the valence shell creating an electron-hole pair.
  - A hole doesn't carry any charge, but it is given a positive sign (+).
- Electron-hole pair: a missing free electron (−) produces a hole (+).

Example: Think of vehicles in a nearly full parking lot:
  - A vehicle ≈ a free electron
  - A parking lot ≈ a hole

Intrinsic and extrinsic semiconductor

- Intrinsic (pure) semiconductor: a pure semiconductor material with a number of holes and electrons are equal, they do not conduct current.
  - Intrinsic semiconductor is a predominating semiconductor.
  - The number of free electrons and number of holes in the valence shell are exactly equal.
  - Examples of intrinsic semiconductor: silicon and germanium.
    They have four valence electrons.
- Extrinsic (impure) semiconductor: a semiconductor material with an added impurity semiconductor.
  - An extrinsic semiconductor is an impure semiconductor formed by adding an impurity to a pure semiconductor.
  - The number of free electrons and number of holes in the valence shell are not equal.

N-type and P-type semiconductors

- A P-type semiconductor is made by adding trivalent impurity atoms to a pure semiconductor (creating holes). It has more holes than electrons.
  - A trivalent impurity atom has three valence electrons.
    Examples of trivalent impurity atoms: gallium (G), aluminium (Al), barium (B), etc.
  - In P-type semiconductors

    $\begin{cases} \text{the majority carriers: holes} \\ \text{the minority carriers: free electrons} \end{cases}$

- An N-type semiconductor is made by adding pentavalent impurity atoms to a pure semiconductor (providing free electrons). It has more free electrons than holes.
  - A pentavalent impurity atom has five valence electrons.
    Examples of pentavalent impurity atoms: phosphorus (P), arsenic (As), bismuth (Bi), etc.
  - In N-type semiconductors:

    $\begin{cases} \text{the majority carriers: free electrons} \\ \text{the minority carriers: holes} \end{cases}$

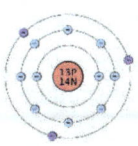

Figure 1.8   Aluminium

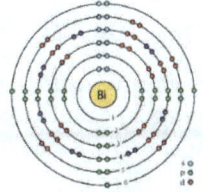

Figure 1.9   Bismuth

Doping

- The process of adding impurities to an intrinsic semiconductor to control its conductivity.
- As doping increases (number of impurity increase), the conductivity of a semiconductor also increases.
- Adding impurity in P-type semiconductors increases more holes.
- Adding impurity in N-type semiconductors increases more free electrons.

1.1.3.   Depletion layer

P–N junction and depletion layer

- P–N junction: when two semiconductor materials (N-type and P-type) are joined together, a P–N junction is formed between the two regions.
  A P–N junction is the border between N-type and P-type materials.
- Depletion layer: a region adjacent to the P–N junction containing no mobile charge carries.
  - The depletion layer is also called the depletion region, junction region, depletion zone, space charge region, space charge layer, etc.
  - The depletion layer acts like a barrier that opposes the flow of electrons from N-side and holes from P-side.

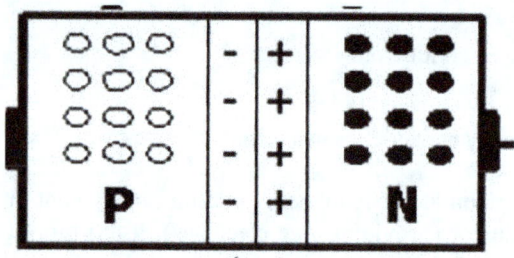

Figure 1.10   Depletion layer

The depletion layer is formed by ionization

- When the P–N junction is formed, the free electrons in the N side will diffuse (spread) across the junction, into the P side and holes in the P side will diffuse toward the N side.

  - The N side ——free electrons diffuse into——▶ the P side
  - The N side ◀——holes diffuse into—— the P side

- Once an electron enters the P side, it combines with a hole within the P side, thus leaving a positive charge in the N side (+ion) and similarly leaving a negative charge in the P side (– ion).

  - N side: positive charge (+ ion) left.
  - P side: negative charge (– ion) left.

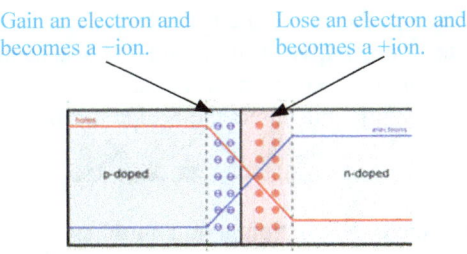

Gain an electron and
becomes a −ion.

Lose an electron and
becomes a +ion.

Figure 1.11   Ionization

A potential barrier created in a P–N junction

- The depletion layer acts as a barrier to the further movement of electrons and holes across the junction.
- The negative ions repel electrons from the N side of the junction.
- The positive ions repel holes from the P side of the junction.

1.2   Bias

1.2.1   Forward and reverse bias

Barrier potential

- Barrier potential (V_B) of a P–N junction: the inherent internal voltage (built-in junction potential) across the depletion layer.
  - Because of the positive and negative ions in the depletion layer, an electric field formed in the depletion layer (it acts as a barrier).
  - Barrier potential is a region in which charges (electrons and holes) are stopped by an obstructive force.
- Barrier potential of a P–N junction depends on the type of semiconductor material and temperature.
  - The typical barrier potential is approximately 0.7 V for silicon and 0.3 V for germanium at 25°C.
  - Silicon (Si): $V_B \approx 0.7$ V,   Germanium (Ge): $V_B \approx 0.3$ V
- Biasing voltage: a DC voltage applied to a P–N junction to make it either conduct or block current.
  - Although the barrier potential (V_B) is an internal potential difference that cannot be measured directly, its effect can be overcome by applying an external voltage – a bias voltage.
  - An external energy (or external voltage) required to overcome the barrier (move the charges through the electric field).

Forward bias

- Forward bias ($+V_{bias}$): when the external voltage is applied across the P–N junction in such a way that allows current pass through a junction.

- In a forward bias setup, the positive terminal of the DC power supply is connected to the P side, and negative terminal of battery is connected to the N side.
  - At the N side of the P–N junction: an electron leaves the negative terminal of the DC power supply and enters the N side to fill a hole at the junction.

$$\text{electrons} \xrightarrow{\text{fill}} \text{holes}$$

  - At the P side of the P–N junction: a hole in the positive terminal of the DC power supply combines with an electron on the N side of the junction.

$$\text{holes} \xrightarrow{\text{combine}} \text{electrons}$$

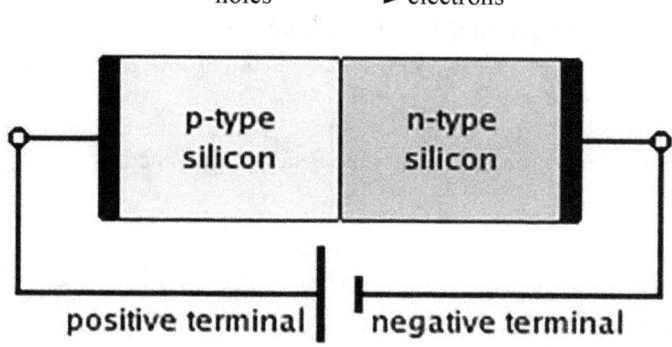

Figure 1.12   Forward bias

The effect of forward biasing on the depletion layer

- When a forward bias voltage ($+V_{bias}$) is applied to the P–N junction: electrons on the N side are pushed and entered the depletion region, the positive ion is neutralized, and the width of the region is reduced. Similarly for holes in the P side. Then the P–N junction is in turn on condition, i.e. the depletion layer is diminished.
  - The positive side of $+V_{bias}$ pushes holes.
  - The negative side of $+V_{bias}$ pushes electrons.
- The width of the depletion layer decreases with an increase in the forward bias voltage ($+V_{bias}$).

$$+V_{bias} \uparrow \rightarrow \text{depletion layer} \downarrow$$

- The width of the depletion layer increases with a decrease in the forward bias voltage.

$$+V_{bias} \downarrow \rightarrow \text{depletion layer} \uparrow$$

- When the forward bias voltage ($+V_{bias}$) is equal to the barrier potential voltage (V_B), the depletion layer disappeared, the current I will increase greatly.

$$+V_{bias} = V_B \rightarrow \text{depletion layer disappeared} \rightarrow I \uparrow\uparrow$$

Reverse bias

- Reverse bias ($-V_{bias}$): when the external voltage is applied across the P–N junction in such a way that blocks the current through a junction.
- In a reverse bias setup, the negative terminal of the battery is connected to the P side, and the positive terminal of the battery is connected to the N side.
- At the P side of the P–N junction: an electron leaves the negative terminal of the DC power supply enters the P side to fill a hole at the junction, leaving more positive ions.

    Electrons $\xrightarrow{\text{fill}}$ holes

- At the N side of the P–N junction: a hole from the positive terminal of the DC power supply enters the N side to combine with an electron at the junction, leaving more negative ions. The width of the depletion layer is increased.

    holes $\xrightarrow{\text{combine}}$ electrons

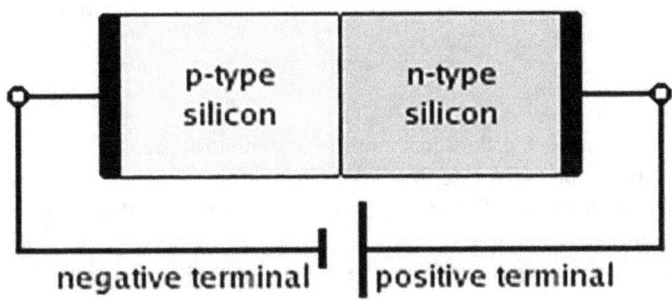

Figure 1.13   Reverse bias

1.2.2   Reverse breakdown

The effect of reverse bias on the depletion layer

- The width of the depletion layer decreases with decrease in the reverse bias voltage ($-V_{bias}$).

    $-V_{bias} \downarrow \rightarrow$ depletion layer $\downarrow$

- The width of the depletion layer increases with increase in the reverse voltage ($-V_{bias}$), and the current would be greatly decreased.

    $-V_{bias} \uparrow \rightarrow$ depletion layer $\uparrow \rightarrow I \downarrow\downarrow$

Forward bias vs. reverse bias

- Forward bias: the negative voltage on the N side promotes the diffusion of electrons by decreasing the width of depletion layer (higher current).

Forward bias permits majority carrier current (electrons in N-type material or holes in P-type material) through the P–N junction.

- Reverse bias: the positive voltage on the N side reduces the diffusion of electrons by increasing the width of depletion layer (lower current).

Reverse bias prevents majority carrier current through the P–N junction.

Majority carriers and minority carriers

- Majority carriers: electrons in N-type semiconductor and holes in P-type semiconductor.
- Minority carriers: holes in N-type semiconductor and electrons in P-type semiconductor.
- Minority carriers provide the small reverse current.
  - The P-type semiconductor has a few minority electrons.
  - The N-type semiconductor has a few minority holes.
- Reverse bias permits minority carrier current and prevents majority carrier current through the P–N junction.
- Forward bias permits majority carrier current and prevents minority carrier current through the P–N junction.

Reverse current (leakage current)

- Reverse current (I_R): a small current in the depletion region because of minority carries (small and can be neglected).
- Reverse current is small and can be neglected because there are few minority carriers in the P–N junction.
  - The negative side of the bias voltage (V_{bias}) pushes electrons into the depletion region.
  - The positive side of the bias voltage (V_{bias}) pushes holes into the depletion region and a small reverse (leakage) current result.

Reverse breakdown

- Reverse breakdown: when a reverse bias voltage V_R is increased to the breakdown voltage, reverse current I_R sharply increases and causes a P–N junction to break down (it conducts in the reverse direction).

$$V_R \uparrow \uparrow \rightarrow \text{P–N junction breaks down} \rightarrow I_R \uparrow \uparrow$$

- Reverse breakdown voltage (V_{BR}): the amount of reverse bias voltage that will make the P–N junction break down and the rapid increase of the current under reverse bias.

The cause of reverse breakdown

- When reverse bias voltage V_R is increased, the high reverse bias voltage exerts a force on the minority electrons at the P side of the P–N junction.
- The minority electrons speed through the P side and collide with other neighboring atoms and knock their valence electrons out of the outmost shell.

- More and more valance electrons are freed; they have enough energy to go through the N side rather than combining with holes in the depletion region.
- This causes a large current to flow, and it increases rapidly with an increase in reverse bias voltage. The result is that the P–N junction is destroyed by the heat that is produced.

  This is known as avalanche breakdown.

$V_R \uparrow\uparrow \rightarrow$ minority electrons accelerate $\rightarrow$ collide with neighboring atoms

$\rightarrow$ knock valence electrons out of outmost shell $\rightarrow$ numbers multiply

$\rightarrow$ avalanche $\rightarrow I_R \uparrow\uparrow \rightarrow$ heat$\uparrow\uparrow \rightarrow$ P–N junction is damaged

1.2.3 I–V characteristic curve of a PN junction

I–V characteristic curve

- The *I–V* characteristic curve of P–N Junction is a graph showing the current flow measured at different given voltages (a nonlinear curve).
- The values of voltage are on the *x*-axis, and the values of current are on the *y*-axis.

  The voltage is independent, and the current is dependent on the voltage.

Figure 1.14   I–V curve

I–V characteristic curve in forward bias

- When the P–N junction is forward biased, forward current (I_F) increases exponentially with increasing forward bias voltage (V_F).
- When a forward bias voltage (V_F) is below the barrier potential (V_B), the forward current (I_F) is very close to zero.

$$V_F < V_B : I_F \approx 0$$

- When the forward voltage equals or exceeds the barrier potential (V_B), the forward current (I_F) sharply increases.

$$V_F \geq V_B : I_F \uparrow\uparrow$$

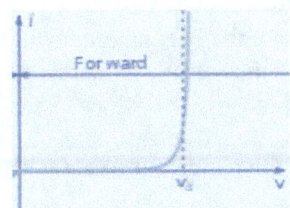

Figure 1.15   Forward bias curve

The barrier potential for silicon and germanium

- Silicon: the barrier potential is approximately 0.7 V for silicon (Si).

  For Si: $V_B \approx 0.7$ V

- Germanium: the barrier potential is approximately 0.3 V for germanium (Ge).

  For Ge: $V_B \approx 0.3$ V

- Most P–N junctions are made of silicon because:
  - less cost (greater abundance).
  - peak inverse voltage of silicon is greater than germanium.
  - withstand excessive heat.
  - has fewer free electrons than germanium at room temperature.

I–V characteristic curve in reverse bias

- When the P–N junction is reverse biased, reverse current (I_R) increases exponentially with an increasing reverse bias voltage (V_R).
- When a reverse bias voltage (V_R) is below the reverse breakdown voltage (V_{BR}), reverse current (I_R) is very close to zero.

  $V_R < V_{BR} : I_R \approx 0$

- When the reverse voltage equals or exceeds the reverse breakdown voltage (V_{BR}), the reverse current ($-I_R$) sharply increases.

  $V_R \geq V_{BR} : -I_R \uparrow\uparrow$

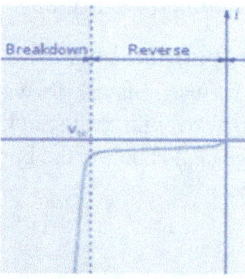

Figure 1.16   Reverse bias curve

Temperature effects on the *I–V* curve

- An increased temperature of the P–N junction will help to increase the velocity of holes and electrons and thus results in increase in the flow of current.
  The above phenomenon applies both to forward and reverse current.
- Temperature and current:
  - Forward bias: increase in temperature causes an increase in forwarding current (I_F).

    temperature $\uparrow \rightarrow I_F \uparrow$

    – Reverse bias: increase in temperature causes an increase in reverse current ($-I_R$).

$$\text{temperature} \uparrow \ \rightarrow \ -I_R \uparrow$$

1.3 Diodes

1.3.1 Introduction to diodes

The diode

- Diode: a two-terminal semiconductor component (a P–N junction device) that conducts current in only one direction.
- Diode structure: a diode is formed when an N-type and a P-type semiconductor material are joined together.
  - The P side of the P–N junction called the anode (positive terminal): P = anode
  - The N side of the P–N junction called the cathode (negative terminal): N = cathode

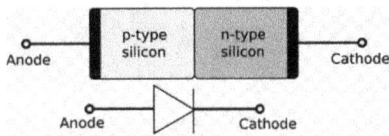

Figure 1.17   Diode

- Diode symbol:

Figure 1.18   Diode symbol

Forward/reverse bias connection

- Forward bias connection: the positive terminal of the DC power supply is connected to the anode, and the negative terminal is connected to the cathode.

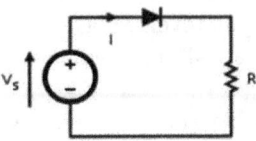

Figure 1.19   Forward bias connection

- Reverse bias connection: the negative terminal of the DC power supply is connected to the anode, and the positive terminal is connected to the cathode.

Types of diodes
- The most common types of diodes are shown in Figure 1.20.
- The cathode of the diode is usually indicated on a real diode by a wide band, a tab, etc.

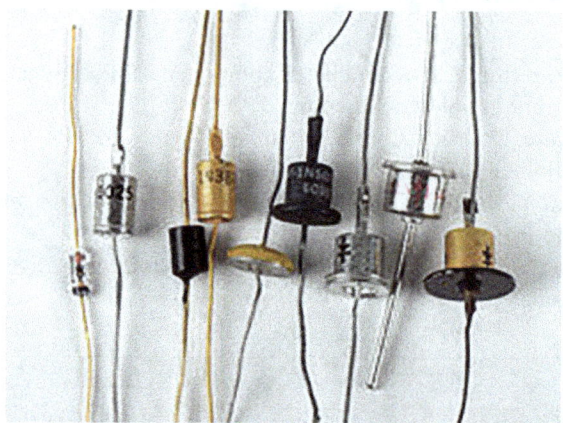

Figure 1.20   Types of diodes

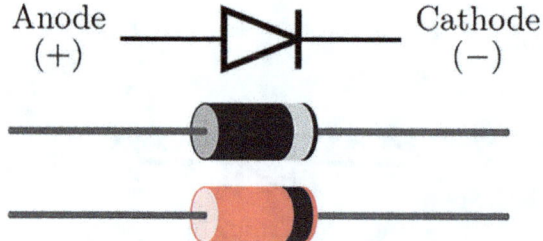

Figure 1.21   The cathode of the diode

1.3.2   Ideal diodes

Ideal diodes
- Ideal diode: a diode that conducts current immediately in forwarding bias and no conduction immediately in reverse bias.
- An ideal diode is a theoretical approximation diode used to solve practical problems.
- An ideal diode functions like an intelligent switch that closes to (on) allow current I_F flow when in the forward bias, when reverse biased, it acts like an open switch (off) to prevent current $(-I_R)$ flow.

- An ideal diode ≈ a switch
  - Forward bias (FB) → switch on → I_F ↑
  - Reverse bias (RB) → switch off → $-I_R$ ↓ ≈ 0

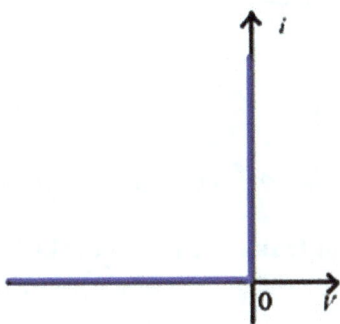

Figure 1.22   An ideal diode

Characteristics of the ideal diode

- When forward biased (switch on): the ideal diode has infinite forward current I_F. It implies that it possesses zero resistance and zero voltage.

  FB → switch on → I_F ↑↑ ≈ ∞ → R_F ≈ 0, $V_F = 0$ (closed circuit) $R \downarrow = \dfrac{V}{I \uparrow}$

- When reverse biased (switch off): the ideal diode has zero current. It implies that it possesses infinite resistance and infinite voltage.

  RB → switch off → $-I_R$ ↓↓ ≈ 0 → R_R ≈ ∞, $V_F = ∞$ (open circuit) $I \downarrow = \dfrac{V}{R \uparrow}$

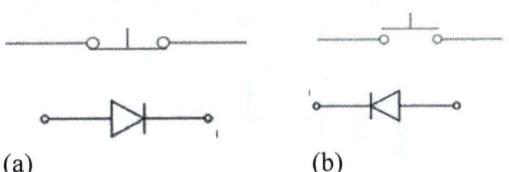

(a)   (b)

Figure 1.23   (a) Switch on (b) Switch off

Open and closed circuits

- An ideal diode acts as a closed circuit when it is forward biased (the "on" state of the diode).
- An ideal diode acts as an open circuit when it is reverse biased (the "off" state of the diode).

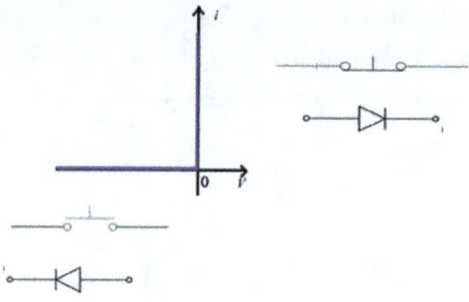

Figure 1.24   Open and closed circuits

Breakdown voltage and leakage current of the ideal diode

- Breakdown voltage: an ideal diode does not have a reverse breakdown voltage.
  The ideal diode does not conduct during reverse bias, so has no possibility of breaking down.
- Reverse leakage current: an ideal diode has zero reverse leakage current in reverse bias condition.
  The ideal diode does not conduct during reverse bias.

1.3.3   Real (practical) diodes

The real diode

- When forward biased: a real diode still acts as a switch when forward biased, but the voltage required to operate this switch is a forward voltage V_F – barrier potential (Si: $V_F \approx 0.7$ V; Ge: $V_F \approx 0.3$ V).
  No current will flow until the voltage applied to the diode exceeds its barrier potential – threshold voltage (0.7 V for Si and 0.3 V for Ge).
- When reverse biased: a real diode is still considered to be an open switch when reverse biased but has a small reverse leakage current.
  Reverse current (I_R): a small current in the depletion region because of minority carries current (it is small and can be neglected).

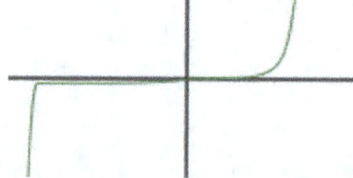

Figure 1.25   I–V curve of a real diode

Breakdown voltage and leakage current of the real diode

- Breakdown voltage: a real diode has a reverse breakdown voltage.
- Reverse leakage current: the real diode has a small reverse leakage current in reverse bias condition.

Real diodes vs. ideal diodes

- Ideal diode
  - An ideal diode acts like a perfect conductor (short circuit – has zero voltage drops across its junction) when it is forward biased.
  - An ideal diode acts like a perfect insulator (open circuit – draws no current) when it is reverse biased.
- Real (practical) diode
  - A practical diode cannot act like a perfect conductor and a perfect insulator.
  - There is a small voltage drop (barrier potential) across the diode when it is forward biased and draws a very small leakage current when it is reverse biased.
  - The small leakage current increases with temperature and reverse voltage.

Table 1.2   Real diodes vs. ideal diodes

Real (practical) diode	Ideal diode
It cannot act as a perfect conductor/insulator.	It acts as a perfect conductor/insulator.
It has small leakage current when reverse biased.	It has no leakage current when reverse biased.
It has low voltage drops across its junction when forward biased (Si ≈ 0.7 V; Ge ≈ 0.3 V).	It has zero voltage drops across its junction when forward biased.
It can be manufactured.	It cannot be made.

Summary

Atomic structure

Table 1.3   Particle

Particle name	Symbol	Charge	Mass	Location
Proton	p^+	Positive +	1 amu	Nucleus
Electron	e^-	Negative −	0	Orbitals
Neutron	n	Neutral	1 amu	Nucleus

Ion

- Ion: an atom that has gained or lost electron(s).
- Positive ion: a neutral atom *loses* an electron.
- Negative ion: a neutral atom *gains* an electron.

Bohr diagram (Bohr model)

- The nucleus is at the center, and the electrons are located on different shells.
- Valence electrons (or free electrons): the electrons in the outmost shell (or valence shell) of an atom.

Arrangements of electrons

- Electrons are located in shells according to rules:
  - The first shell can only hold two electrons.
  - The second shell can hold up to (maximum capacity) eight electrons.
  - The third shell can hold up to 18 electrons.
  - The fourth shell can hold up to 32 electrons.
  - ... ...
  - The maximum number of electrons in the outmost shell (valence shell) is 8.
- Octet rule: the tendency of atoms to have eight electrons in the valence shell (outmost shell).
- Atoms that do not have eight electrons in their outer shell will tend to gain or lose electrons.

Semiconductor

- Semiconductor: a material with electrical conductivity between that of an insulator and a conductor.
- The elements silicon (Si) and germanium (Ge) are semiconductors and have four valence electrons which can be shared with neighboring atoms.

Electron-hole pair

- Hole (+): a hole corresponds to the vacancy created by a free electron in a semiconductor.
- Electron-hole pair: a missing free electron (−) produces a hole (+).

Intrinsic and extrinsic semiconductor

- Intrinsic (pure) semiconductor: a pure semiconductor material with a number of holes and electrons are equal, they do not conduct current (such as silicon and germanium).
- Extrinsic (impure) semiconductor: a semiconductor material with an added impurity semiconductor.

N-type and P-type semiconductors

- P-type semiconductor is made by adding trivalent impurity atoms to a pure semiconductor, It has more holes than electrons.
- N-type semiconductor is made by adding pentavalent impurity atoms to a pure semiconductor. It has more free electrons than holes.
- Doping: the process of adding impurities to an intrinsic semiconductor to control its conductivity.

P–N junction and depletion layer

- P–N junction: when two semiconductor materials (N-type and P-type) are joined together, a P–N junction is formed between the two regions.
- Depletion layer: a region adjacent to the P–N junction containing no mobile charge carries.

- The depletion layer acts as a barrier to the further movement of electrons and holes across the junction.
- The negative ions repel electrons from the N side of the junction.
- The positive ions repel holes from the P side of the junction.

Barrier potential
- Barrier potential (V_B) of a P–N junction: the inherent internal voltage across the depletion layer.

$$\text{Silicon (Si): } V_B \approx 0.7 \text{ V} \qquad \text{Germanium (Ge): } V_B \approx 0.3 \text{ V}$$

- Biasing voltage: a DC voltage applied to a P–N junction to make it either conduct or block current.

Forward bias
- Forward bias ($+V_{bias}$): when the external voltage is applied across the P–N junction in such a way that allows current pass through a junction.
- In a forward bias setup, the positive terminal of the DC power supply is connected to the P side, and negative terminal of battery is connected to the N side.
- $+V_{bias} \uparrow \rightarrow$ depletion layer $\downarrow$
- $+V_{bias} \downarrow \rightarrow$ depletion layer $\uparrow$
- $+V_{bias} = V_B \rightarrow$ depletion layer disappeared $\rightarrow I \uparrow\uparrow$

Reverse bias
- Reverse bias ($-V_{bias}$): when the external voltage is applied across the P–N junction in such a way that blocks the current through a junction.
- In a reverse bias setup, the negative terminal of the battery is connected to the P side, and the positive terminal of the battery is connected to the N side.
- $-V_{bias} \downarrow \rightarrow$ depletion layer $\downarrow$
- $-V_{bias} \uparrow \rightarrow$ depletion layer $\uparrow \rightarrow I \downarrow\downarrow$

Majority carriers and minority carriers
- Majority carriers: electrons in N-type semiconductor and holes in P-type semiconductor.
- Minority carriers: holes in N-type semiconductor and electrons in P-type semiconductor.
- Minority carriers provide the small reverse current.
- Reverse bias permits minority carrier current and prevents majority carrier current through the P–N junction.
- Forward bias permits majority carrier current and prevents minority carrier current through the P–N junction.

Reverse current (leakage current)
- Reverse current (I_R): a small current in the depletion region because of minority carries.
- Reverse current is small and can be neglected.

Reverse breakdown

- Reverse breakdown:

$$V_R \uparrow \; \rightarrow \; \text{P–N junction breaks down} \rightarrow I_R \uparrow\uparrow$$

- Reverse breakdown voltage (V_{BR}): the amount of reverse bias voltage that will make the P–N junction break down and the rapid increase of the current under reverse bias.

I–V characteristic curve

- The *I–V* characteristic curve of P–N Junction:

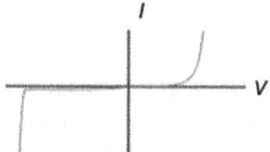

Figure 1.26   I–V curve

I–V characteristic curve in forward bias

- When $V_F < V_B : I_F \approx 0$
- When $V_F \geq V_B : I_F \uparrow\uparrow$

I–V characteristic curve in reverse bias

- When $V_R < V_{BR} : I_R \approx 0$
- When $V_R \geq V_{BR} : -I_R \uparrow\uparrow$

Temperature effects on the I–V curve

- Forward bias: temperature $\uparrow \; \rightarrow \; I_F \uparrow$
- Reverse bias: temperature $\uparrow \; \rightarrow \; -I_R \uparrow$

The diode

- Diode: a two-terminal semiconductor component (a P–N junction device) that conducts current in only one direction.
- Diode symbol:

Figure 1.27   Diode symbol

Table 1.4   Real diodes vs. ideal diodes

Real (practical) diode	Ideal diode
It cannot act as a perfect conductor/insulator.	It acts as a perfect conductor/insulator.
It has small leakage current when reverse biased.	It has no leakage current when reverse biased.
It has low voltage drops across its junction when forward biased (Si ≈ 0.7 V; Ge ≈ 0.3 V).	It has zero voltage drops across its junction when forward biased.
It can be manufactured.	It cannot be made.

Self-test

1.1 **1.** A (   ) is a positively charged particle.

2. In an ion: No. of electrons ≠ No. of (   )

3. Atoms that do not have (   ) electrons in their outer shell will tend to gain or lose electron(s).

4. The third shell can hold up to (   ) electrons.

5. The elements silicon and germanium have (   ) valence electrons which can be shared with neighboring atoms.

6. An N-type semiconductor is made by adding pentavalent impurity atoms to a pure (   ). It has more free electrons than holes.

7. The barrier potential of a P–N junction is the inherent internal voltage across the (   ) layer.

1.2 **8.** The biasing voltage is a (   ) voltage applied to a P–N junction to make it either conduct or block current.

9. The width of the layer decreases with an (   ) in the forward bias voltage.

10. In a (   ) bias setup, the positive terminal of the DC power supply is connected to the P side, and negative terminal of battery is connected to the N side.

11. The barrier potential is approximately (   ) V for germanium.

12. Increase in temperature causes an (   ) in reverse current I_R.

1.3 **13.** The N side of the P–N junction is called the (   ).

14. Reverse bias connection: the negative terminal of the DC power supply is connected to the anode, and the positive terminal is connected to the (   ).

15. The (   ) of the diode is usually indicated on a real diode by a wide band, a tab, etc.

16. An ideal diode acts as an open circuit when it is (   ) biased.

17. There is a small voltage drop (barrier potential) across the diode when it is (   ) biased.

Chapter 2

Types of diodes and their applications

Chapter outline

2.1 Zener diode and its applications

2.1.1 Diodes classification

Types of diodes

- Recall – diode: a two-terminal semiconductor component (a P–N junction device) that conducts current in only one direction.
- Types of diodes: there are different types of diodes that are available for use in electronics. Diodes are classified according to their characteristics. Following are some of the types of diodes:
  - Zener diode
  - Light-emitting diode (LED)
  - Photodiode
  - Varicap (varactor) diode
  - Laser diode
  - Schottky diode
  - Gunn diode
  - PIN diode
  - BARITT diode
  - Tunnel diode
  - ...

Overview of different types of diodes

- Zener diode: a diode designed in such a way that it can work in the reverse breakdown region (a highly doped diode).

Figure 2.1 (a)   Zener diode symbol    *Figure 2.1 (b)   A Zener diode*

- LED: a diode that emits visible light when current passes through it. It operates in forwarding bias condition.

Figure 2.2   LED

- Photodiode: a light-sensitive diode that conducts current when a certain amount of light falls on it. It operates in reverse bias condition.

Figure 2.3   Photodiode

- Varicap diode (varactor diode): a diode whose internal capacitance varies with the variation of the reverse voltage. (It acts as a variable capacitor for giving the preferred capacitance changes.)

Figure 2.4   Varicap diode

- Laser diode (injection laser or diode laser): a diode that generates laser light (coherent light) of high intensity.
  Coherent light: a beam of photons that have the same frequency and phase (laser beam is coherent).

Figure 2.5   Laser diode

- Schottky diode (Schottky barrier diode): a diode that has a very fast switching time and a low forward voltage drop.

Figure 2.6   Schottky diode

- BARITT diode (barrier injected transit-time diode): a high-frequency diode used for generating microwave signals. It uses thermionic emission (minority carrier injection from forward-biased (FB) junctions) rather than avalanche multiplication.
  The BARITT diode consists of two back-to-back diodes.

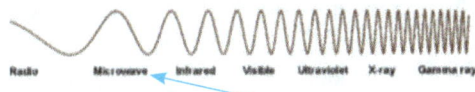

Figure 2.7   Microwave

- Gunn diode: a high-frequency semiconductor component used for generating microwave signals. It has two terminals and formed by only N-type material (not a diode in the form of a P–N junction).

Figure 2.8 (a)   Gunn diode   *Figure 2.8 (b)   Gunn diode symbol*

- PIN diode: a special diode that has a layer of intrinsic semiconductor (no doping) that is sandwiched between a P-and an N-type semiconductor material.

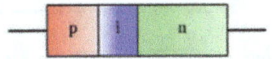

Figure 2.9   PIN diode

- Tunnel diode: a heavily doped diode that exhibits negative resistance, meaning the current decreases as the voltage increases. It works on the principle of the tunneling effect.
  Tunneling effect: the ability of electrons to pass a barrier even though usually shouldn't be able to pass.

   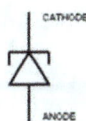

Figure 2.10 (a)   Tunnel diode   *Figure 2.10 (b)   Tunnel diode symbol*

2.1.2   Zener diodes

Introduction to Zener diode

- Zener diode: a diode designed in such a way that it can work in the reverse break down region (a highly doped diode).

  Clarence Melvin Zener (1905–1993): an American physicist who discovered the Zener effect (a breakdown occurs in a reverse-biased P–N junction).
- Normal diode vs. Zener diode:
  - Normal diodes cannot operate in the reverse breakdown region, the diode would be destroyed if the reverse voltage V_R equals or exceeds the reverse breakdown voltage V_{BR} (the reverse current I_R will be sharply increased).

    $V_R \geq V_{BR} \rightarrow I_R \uparrow\uparrow$
  - Zener diode is designed to work in the reverse breakdown region without damage it.
- The symbol of the Zener diode:
  - The symbol for a regular diode has a straight line representing the cathode.
  - The symbol for a Zener diode has a bent line that represents the letter Z (for Zener).

The reverse bias of a Zener diode

- The reverse bias of a Zener diode: the negative terminal of the power supply is connected to the anode of the Zener diode, and the positive terminal is connected to the cathode.

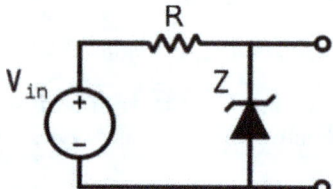

Figure 2.11   Zener diode circuit

- When the Zener voltage V_Z is exceeded and the Zener diode is operating in the Zener region, the voltage across the Zener diode stays constant even with the changing current.

  In the Zener region: $I_R \rightarrow \vec{V}_z$

The Zener breakdown

- The avalanche breakdown occurs because of the ionization of free electrons and hole pairs.

  $V_R \uparrow\uparrow \longrightarrow$ minority electrons accelerate $\longrightarrow$ collide with neighboring atoms $\longrightarrow$ knock valence electrons out of outermost shell $\longrightarrow$ numbers multiply $\longrightarrow$ avalanche $\longrightarrow I_R \uparrow\uparrow$

- The Zener breakdown occurs because of the heavy doping in the P–N junction. Recall doping: the process of adding impurities to an intrinsic semiconductor to control its conductivity.
  - A heavily doped P–N junction diode has a much narrower depletion layer.

    Heavy doping $\rightarrow$ narrow depletion layer $\rightarrow I_R \uparrow\uparrow$

  - As a result, it takes little reverse bias voltage to cause the diode to go into breakdown.

Zener voltage

- Zener voltage (V_Z): the amount of reverse bias voltage that will make the P–N junction break down and allow the Zener diode to conduct in the reverse direction.
- Zener breakdown occurs with low Zener voltage V_Z ratings ($V_Z < 5V$).
- Avalanche breakdown occurs at a higher value of Zener voltage V_Z than does Zener breakdown ($V_Z > 5V$).

Breakdown characteristics

- The Zener knee current (I_{ZK}) or minimum current ($I_{Z(min)}$) is the turning point (knee point) of the Zener diode into a breakdown. It is the minimum reverse current required to maintain constant breakdown voltage.

    I_Z must $> I_{ZK}$ (or $I_{Z(min)}$)

    At the knee point, the breakdown effect begins.

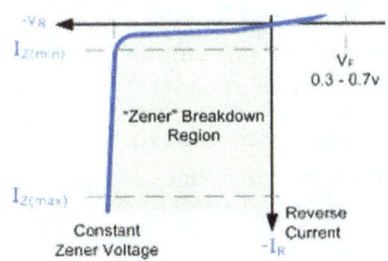

Figure 2.12   Zener diode circuit

- Maximum Zener current ($I_{Z(Max)}$): the maximum allowable amount of current that can pass through the Zener diode without damaging it.
- Zener test current (I_{ZT}): the current value to ensure good Zener voltage V_Z regulation. V_Z is determined at this current level.

Zener regulation

Zener diodes are mainly used as a voltage regulator to produce a stable output voltage.

- The Zener voltage V_Z is approximately the constant when current I_Z is between the maximum Zener current I_{ZM} and the Zener knee current I_{ZK}.

$$I_{ZK} \le I_Z \le I_{ZM} \rightarrow V_Z \approx \text{constant}$$

- The Zener diode continues to regulate the voltage until I_Z reaches the Zener knee current I_{ZK} (or $I_{Z(min)}$). If $I_Z < I_{Zk} \rightarrow$ out of regulation.

Equivalent circuit for Zener diode

- Zener equivalent circuit: a Zener diode acts like a battery in the breakdown region.
- Equivalent circuit of an ideal Zener diode: an ideal Zener diode can be replaced by a voltage source.
- Equivalent circuit of a practical Zener diode: an actual Zener diode can be replaced by a voltage source in series with a resistor.

Figure 2.13 (a)   Ideal Zener diode   Figure 2.13 (b)   Practical Zener diode

- Zener impedance: the equivalent series impedance of a practical Zener diode.

$$Z_Z = \frac{\Delta V_Z}{\Delta I_Z} = \frac{\Delta V_Z}{I_{ZT} - I_{Zk}} \qquad I_{ZT} - \text{Zener test current}; I_{Zk} - \text{Zener knee current}$$

- ΔV_Z: a small change of Zener voltage.
- ΔI_Z: a small change of Zener current.

Example: Determine the Zener impedance of a practical Zener diode if the Zener voltage changes from 35 mV to 90 mV, and the Zener current changes from 9 mA (I_{ZK}) to 14 mA (I_{ZT}).

$$Z_Z = \frac{\Delta V_Z}{I_{ZT} - I_{Zk}} = \frac{(90-35)\text{mV}}{(14-9)\text{mA}} = \frac{55\text{ mV}}{5\text{ mA}} = 11\,\Omega$$

Example: A Zener diode has a resistor of 6 Ω. Determine the voltage across the Zener diode (Zener terminals) when the current is 18 mA and the voltage V_Z is 9 V.

$$V = 9\text{ V} + (18\text{ mA})(6\,\Omega)$$

$$= 9\text{ V} + 108\text{ mV} \qquad \text{Milli: } 10^{-3}$$

$$\approx 117\text{ V}$$

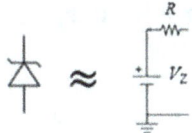

Figure 2.14   Practical Zener diode – an example

2.1.3   Zener diode as a voltage shifter and a clipper

Main applications of Zener diodes:
- Voltage regulator
- Clipper
- Voltage shifter
- ...

Overview of different applications of Zener diodes
- Voltage shifter (level shifter): a circuit used to convert voltage from one level to another (allowing different voltage requirements).
- Clipper (Zener limiter): a circuit that can clip, limit, or cut off an input waveform to prevent the output waveform beyond a determined value.
- Voltage regulator: a circuit is used to regulate the output voltage level into a lower, fixed level that remains constant for any changes in load or input voltage.

Zener diode as a voltage shifter
- Voltage shifter (level shifter): a circuit used to convert voltage from one level to another.
- A Zener diode with a resistor can act as a voltage shifter. It has the ability to maintain a lower steady output voltage to the Zener diode's breakdown voltage.

Example: A typical Zener voltage shifter circuit (from 5 V to 3.3 V voltage level-shifting).
– Use a Zener diode that has a breakdown voltage of 3.3 V.
– If the input voltage V_{in} for the circuit is 5 V, then the output voltage V_{out} for this circuit will be 3.3 V.

$$V_{in} = 5 \text{ V}$$

$$V_{out} \approx 3.3 \text{ V (Zener breakdown voltage)}$$

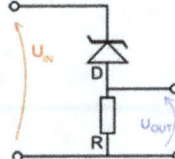

Figure 2.15   Zener voltage shifter circuit

Zener diode as a clipper (Zener limiter)

- Introduction to the clipper
  - Clipper (Zener limiter): a circuit that can clip, limit, or cut off an input waveform to prevent the output waveform beyond a determined value.
  - A clipper is the circuit that can limit positive or negative (or both) amplitude of the input waveform to the desired output level.
  - A single Zener diode can limit one side of a sinusoidal waveform to the Zener voltage while clamping the other side to near zero (0.7 V for silicon or 0.3 V for germanium).
  - With two Zener diodes in series (facing each other), the output waveform can be limited to the Zener voltage on both polarities.
    Waveform clippers can also be used to prevent voltage spikes from affecting circuits that are connected to the power supply.
- Three ways to limit waveform
  - Positive limiter: the positive peak of input sinusoidal voltage will be limited to the selected output Zener voltage.

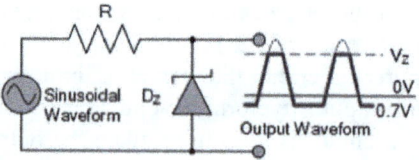

Figure 2.16   Positive limiter

(+) cycle: the output voltage V_{out} equals the Zener voltage V_z.

$$V_{out} = V_z$$

(−) cycle: the Zener diode acts as an FB diode and the output voltage

V_{out} is −0.7 V for silicon and −0.3 V for germanium.

$V_{out} = -0.7$ V (Si) or $V_{out} = -0.3$ V (Ge)   Zener diode = FB diode

  - Negative limiter: the negative peak of input sinusoidal voltage will be limited.

Figure 2.17   Negative limiter

(+) cycle: the output voltage V_{out} is 0.7 V for silicon and 0.3 V for germanium.

$V_{out} = 0.7$ V (Si) or $V_{out} = 0.3$ V (Ge)   Zener diode = FB diode

($-$) cycle: the output voltage V_{out} equals the negative Zener voltage ($-V_Z$).

$$V_{\text{out}} = -V_Z$$

—   Back-to-back Zener limiter: with two opposing Zener diodes, the input sinusoidal waveform can be clipped to the Zener voltage on both polarities.

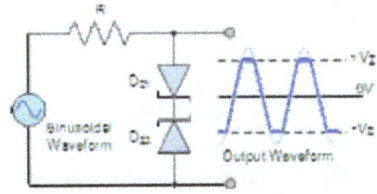

Figure 2.18   Back-to-back Zener limiter

(+) cycle:

    ○   Zener diode D_1: $V_{\text{out1}} = 0.7\text{V}$ (Si) or $V_{\text{out1}} = 0.3\text{V}$ (Ge)
        Zener diode = FB diode

    ○   Zener diode D_2: the output voltage V_{out} equals the Zener voltage V_Z.

$$V_{\text{out2}} = V_{Z2}$$

    ○   Total output voltage: $V_{\text{out}} = V_Z + 0.7$ V (Si) or $V_{\text{out}} = V_Z + 0.3$ V (Ge)

($-$) cycle:

    ○   Zener diode D_1: $V_{\text{out1}} = -0.7$ V (Si) or $V_{\text{out1}} = -0.3$ V
        Zener diode = FB diode

    ○   Zener diode D_2: the output voltage V_{out} equals the negative Zener voltage V_Z.

$$V_{\text{out2}} = -V_{Z2}$$

    ○   Total output voltage: $V_{\text{out}} = -V_Z + (-0.7 \text{ V}) = -V_Z - 0.7$ V (Si)

$$V_{\text{out}} = -V_Z + (-0.3 \text{ V}) = -V_Z - 0.3 \text{ V (Ge)}$$

2.1.4   *Zener diode as a voltage regulator*

Introduction to voltage regulator

*   Voltage regulator: a device designed to automatically maintain a stable output voltage despite variations in the input signal or load.
*   Zener diodes are widely used as voltage regulators because of their reverse bias characteristics. In the reverse breakdown region, the voltage across the Zener diode (V_Z) is close to constant over a wide range of currents.
*   Zener voltage regulator circuit:
    Because of the low impedance of the Zener diode, a resistor R is used to limit the current through the circuit.

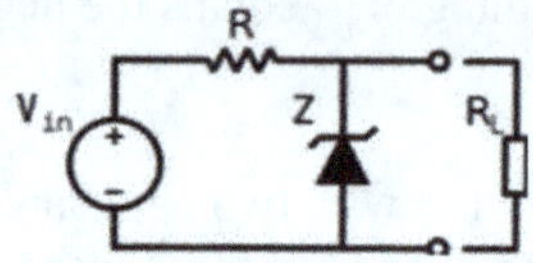

Figure 2.19   Voltage regulator

Line regulation

- Line regulation: the ability of a voltage regulator to maintain the output voltage V_{out} level with a varying input voltage V_{in}.
- If the input voltage varies, the current in the circuit will change, consequently the voltage across the resistor will change, but the constant Zener diode voltage will maintain the output voltage V_{out} at a constant level.

$$V_{in} \to I \to IR \to \overrightarrow{V_z} = \overrightarrow{V_{out}}$$

- If the input voltage increases: $V_{in} \uparrow \to I \uparrow \to I \uparrow R \to \overrightarrow{V_z} = \overrightarrow{V_{out}}$
- If the input voltage decreases: $V_{in} \downarrow \to I \downarrow \to I \downarrow R \to \overrightarrow{V_z} = \overrightarrow{V_{out}}$

Example: A voltage regulator in Figure 2.19 with a 5.6 V Zener diode at a Zener knee current (I_{ZK}) of 1 mA, Zener maximum current (I_{Zm}) of 162 mA, Zener test current (I_{ZT}) of 45 mA, Zener impedance of 5 Ω, and resistor R of 80 Ω. Determine the input voltages range (the minimum and maximum voltage) for the circuit that can be regulated.

- Given: $V_z = 5.6$ V, $I_{ZK} = 1$ mA, $I_{Zm} = 162$ mA, $I_{ZT} = 45$ mA, $Z_z = 5$ Ω, and R = 80 Ω
- Find $V_{in\,(min)}$: $V_{in\,(min)} = I_{ZK} R + V_{out}$

$$V_{in\,(max)}: V_{in\,(max)} = I_{ZM} R + V_{out}$$

I_{ZK} – the Zener knee current (minimum reverse current)

I_{ZM} – the maximum Zener current

$V_{in\,(min)}$ and $V_{in\,(max)}$ – the minimum and maximum input voltage

Solution:

$$V_{out} = V_z - \Delta V_z$$
$$\Delta V_z = (I_{ZT} - I_{ZK}) Z_z, \quad Z_z = \frac{\Delta V_z}{I_{ZT} - I_{Zk}}, \qquad \Delta V_z - \text{a small change of Zener voltage}$$
$$V_{out} = V_z - \Delta V_z = V_z - (I_{ZT} - I_{ZK}) Z_z$$
$$= 5.6\ \text{V} - (45\ \text{mA} - 1\ \text{mA})(5\ \Omega)$$
$$= 5.6\ \text{V} - 220\ \text{mV} = 5{,}600\ \text{mV} - 220\ \text{mV} = 5{,}380\ \text{mV}$$

$$V_{in\,(min)} = I_{ZK}R + V_{out} = (1\text{ mA})\,(80\ \Omega) + 5{,}380\text{ mV} = 5{,}460\text{ mV} = \textcolor{blue}{5.46\text{ V}}$$

Milli: 10^{-3}

$$V_{in\,(max)} = I_{Zm}R + V_{out} = (162\text{ mA})\,(80\ \Omega) + 5{,}380\text{ mV} = 18{,}340\text{ mV} = \textcolor{blue}{18.34\text{ V}}$$

Load regulation

- Load regulation: the ability of a voltage regulator to maintain the output voltage V_{out} level with a varying load.
- If the load resistor varies, the load current will change, consequently the current in the Zener diode branch will change, but the steady Zener diode voltage will maintain the output voltage V_{out} at a constant level.

$$\cancel{R_L} \to \cancel{I_L} \to \cancel{I_Z} \to \overrightarrow{V_Z} = \overrightarrow{V_{out}}$$

- If the load resistance increases: $R_L\uparrow \to I_L\downarrow \to I_Z\uparrow \to \overrightarrow{V_Z} = \overrightarrow{V_{out}}$
- If the load resistance decreases: $R_L\downarrow \to I_L\uparrow \to I_Z\downarrow \to \overrightarrow{V_Z} = \overrightarrow{V_{out}}$

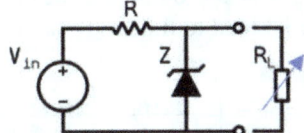

Figure 2.20   Load regulation

Example: The voltage regulator in Figure 2.20 with a 9.1 V Zener diode at a Zener knee current (I_{ZK}) of 0.5 mA. The resistor R is 100 Ω, and the supply voltage (V_{in}) is 20 V.

a) Determine the maximum current for the circuit that can be regulated.
b) Determine the minimum value of load resistance for the circuit that can be regulated.
   - Given:
     $V_Z = 9.1$ V, $I_{ZK} = 0.5$ mA, $R = 100\ \Omega$, and $V_{in} = 20$ V.
   - Find: $I_{L(max)}$ and $R_{L(min)}$.

Solution:

a) - $I_{L(max)} = I_T - I_{ZK}$

- $I_T = \dfrac{V_{in} - V_Z}{R} = \dfrac{20\text{ V} - 9.1\text{ V}}{100\ \Omega} \approx 0.109\text{ A} = 109\text{ mA}$ Milli: 10^{-3}

- $I_{L(max)} = I_T - I_{ZK} = 109\text{ mA} - 0.5\text{ mA} = 108.5\text{ mA} \approx \textcolor{blue}{0.11\text{ A}}$

I_{ZK} – the Zener knee current (minimum reverse current)

b) $R_{L(min)} = \dfrac{V_z}{I_{L((max)}} = \dfrac{9.1\ \text{V}}{0.11\ \text{A}} \approx 82.7\ \Omega$

R_L cannot less than 82.7 Ω, otherwise the circuit will loss of voltage regulation.

Percent regulation

- Percent regulation can be used to specify the performance of a voltage regulator. It can be in terms of line regulation or load regulation.
- Percent line regulation is expressed as a percent of a small change in the output voltage over a small change in the input voltage.

$$\text{Percent line regulation} = \left(\frac{\Delta V_{output}}{\Delta V_{input}} \right) 100\%$$

- Percent load regulation is not a fixed number but rather presented as a percentage in response to changes at the output. It means over the permissible load range (from minimum or no load to full load) the regulation can change.

$$\text{Percent load regulation} = \left(\frac{V_{No\ load} - V_{Full\ load}}{V_{Full\ load}} \right) 100\%$$

- $V_{No\ load}$: percent voltage with no load (or minimum load).
- $V_{Full\ load}$: percent voltage with full load.

Example: Determine the percent voltage regulation if a voltage regulator with percent voltage with no load 12 V and percent voltage with full load 11.5 V.

- Given: $V_{No\ load} = 12\ \text{V}$, $V_{Full\ load} = 11.5\ \text{V}$

- Find: $\left(\dfrac{V_{No\ load} - V_{Full\ load}}{V_{Full\ load}} \right) 100\%$

- Solution: percent voltage regulation $= \left(\dfrac{V_{No\ load} - V_{Full\ load}}{V_{Full\ load}} \right) 100\%$

$$= \frac{12\ \text{V} - 11.5\ \text{V}}{11.5\ \text{V}} \, 100\% \approx 4.35\ \%$$

The smaller % regulation, the better.

2.2   Special types of diodes

2.2.1   Varicap diode

Introduction to varicap diode

- Varicap diode (varactor diode): a specially designed diode whose junction capacitance varies with the variation of the reverse bias voltage (V_R).
  - It acts as a variable capacitor for giving the preferred capacitance changes.
  - The junction capacitance can be changed by changing the reverse voltage.

- Symbol of a varicap diode: a varactor diode uses a symbol that combines both diode and capacitor.

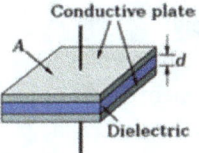

Figure 2.21   Varicap diode

- Equivalent circuit of a varicap diode $\approx$ a variable capacitor

Capacitor

- A capacitor consists of two conductive metal plates separated by an insulating material (the dielectric).

Conductive plate

Dielectric

Figure 2.22   A capacitor

- Factors affecting capacitance (C):
  - the area of plates (A)
  - the distance between the two plates (d)
  - the dielectric constant (k)
  Different dielectric materials have different dielectric constants (k).
- The capacitance (C) of the capacitor is dependent upon the area (A) of the plates and the distance (d) between them. $C = 8.85 \times 10^{-12} \dfrac{kA}{d}$

  $A \rightarrow C$

  $d \rightarrow C$

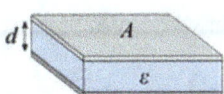

Figure 2.23   Capacitance

Depletion layer and reverse bias voltage

- As with any diode, a varactor diode is formed when N-type and P-type semiconductor materials are joined together. If the reverse voltage is changed so does the size of the depletion layer.

  Depletion layer: a region adjacent to the P–N junction containing no mobile charge carries.

  - The width of the depletion layer decreases on decreasing the reverse-biased voltage.
  - The width of the depletion layer increases on increasing the reverse-biased voltage.
- The varactor diode behaves as a capacitor (insulating depletion layer acting as a dielectric). The N-type and the P-type regions can be considered to be the two metal plates.

The C–V characteristics of varicap diode

- By changing the reverse bias voltage on the varicap diode will change the capacitance.
- The capacitance decreases with the increase in reverse-biased voltage (V_R) and vice versa.

$$V_R \rightarrow \text{depletion layer} \rightarrow \text{capacitance}$$

Varicap diodes vs. variable capacitors

- The smaller size and lightweight.
- Low noise (operate in reverse bias voltage).
- Reliable (a solid-state component – less sensitive to the environment).
- The ability to get different values of capacitances just by changing the voltage applied.

Applications of varicap diodes

- Turning circuit (the tuner in radio receivers).
- Parametric amplifier (varying circuit parameters).
- Frequency/phase modulators: varying the frequency/phase of the signal.
- Voltage-controlled oscillator (the input voltage controls the frequency of oscillations), etc.

2.2.2   Light-emitting diode

Introduction to LED

- LED: a diode that emits light when current passes through it in the forward bias condition.
- The symbol of LED:

Figure 2.24   Light-emitting diode

- Silicon and germanium are not used in LED
  - Silicon and germanium are common materials used to make normal P–N junction diodes.
  - Silicon or germanium is very poor at transmitting light. Instead, they release energy in the form of heat.

Gallium arsenide (GaAs)

- The semiconductor materials used in LED: gallium arsenide (GaAs), gallium arsenide **phosphide** (GaAsP), gallium phosphide (GaP), etc. (They are light-emitting materials.)

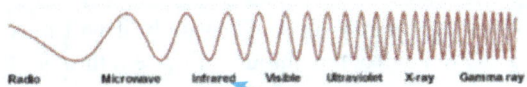

Radio   Microwave   Infrared   Visible   Ultraviolet   X-ray   Gamma ray

Figure 2.25   Infrared light

- GaAs emits infrared light that is invisible.

  Infrared LED can be used in remote controllers for televisions and other household electronics.

Gallium arsenide phosphide (GaAsP) and gallium phosphide (GaP)

- The light-emitting semiconductor material determines the LED's color.
- LEDs produce different colors by using different semiconductor materials which produce photons at different wavelengths or frequencies.

  The visible colors from shortest to longest wavelength or longest to shortest frequency are a rainbow of colors: red, orange, yellow, green, blue, indigo, and violet. Red light has a lower frequency (f) and longer wavelength (λ).

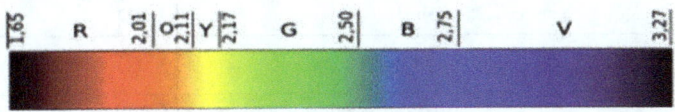

Figure 2.26   Visible colors

- GaAsP emits red or yellow light at different wavelengths.
- GaP emits red, yellow, and green lights at different wavelengths.

Table 2.1   Semiconductor materials

Material	Produced energy/light
Si and Ge	Heat
GaAs	Infrared light (invisible light)
GaAsP	Red or yellow lights
GaP	Red, yellow, and green lights

LED vs. diode

- LED – when a LED is forward biased, energy is released in the form of light.

  Inverse current $I_F \uparrow \rightarrow$ light $\uparrow \rightarrow$ glowing

- Diode – when a diode is forward biased, energy is released in the form of heat energy. It conducts current in one direction.

  Inverse current $I_F \uparrow \rightarrow$ heat $\uparrow$ (not glowing)

- LEDs have higher forward voltage V_F and lower reverse breakdown voltage V_{BR}.
- Use a current-limiting resistor for LED: usually it will have a resistor (R) in series with an LED to control the amount of current through the LED.

$$R = \frac{V_p - V_F}{I_F}$$

where V_p is the peak voltage of the power sauce, V_F is the LED's minimum required forward voltage, and I_F is the forward current.

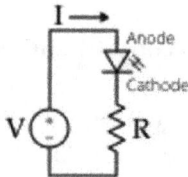

Figure 2.27   An LED circuit

Table 2.2   LED vs. diode

	Diode	LED
Symbol		
Energy	Converts energy into heat	Converts energy into light
Material	Silicon and germanium	GaAs, GaAsP, GaP, etc.
Reverse breakdown voltage V_{BR}	Higher	Lower
Forward voltage V_F	Lower	Higher

LED applications
- Traffic signals
- Camera flashes
- Medical devices
- Seven-segment display (for digital clock, etc.): 0–9
- Multisegment displays: A, B, C, ..., Z
- Christmas lights
- Horticultural grow lights
- Automotive headlamps
- Aviation lighting
- Virtual sky, etc.

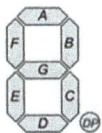

Figure 2.28   Seven-segment display

Figure 2.29   Aviation lighting (Luxair Boeing 737-700)

Figure 2.30   Automotive headlamps

2.2.3   Photodiode

Introduction to photodiode
- Photodiode (photodetector, photo sensor, or light detector): a light-sensitive diode that converts light into electrical current. It is specially designed to operate in reverse bias conditions.
- The symbol of photodiode:

Figure 2.31   Photodiode

Principle of operation
- The operating principle of the photodiode: when the junction of the photodiode is lit up then the current starts flowing through it.
  - When a photon (the fundamental particle of light) of sufficient energy enters the depletion region of a P–N junction, it may strike an atom with sufficient energy to release the valence electron.

    – A photodiode has a small transparent window that allows light to be incident on the P–N junction. The current generated is proportional to the absorbed light intensity.

• The typical material used to make a photodiode: silicon (Si), germanium (Ge), indium gallium arsenide (InGaAs), Gallium Phosphide (GaP), etc.

• Dark current I_D: a small reverse current (unwanted current) that flows through a photodiode with no incident light (no photons are entering the device).

Applications of photodiode

• Cameras
• TVs and remote controls
• Smoke detectors
• Bar code scanners
• Optical communication devices (fiber-optic cable)
• Solar cell panels
• Position sensors
• Automotive devices
• Surveying instruments
• Medical devices, etc.

Figure 2.32   Fiber-optic cable

Figure 2.33   Solar cell panels

Photodiodes vs. LED

• LED: LED converts electrical energy into light energy (it emits light when current passes through it).

    Electrical energy → Light energy

    $I \uparrow$ → Light $\uparrow$

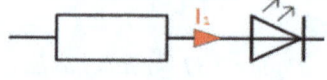

Figure 2.34 $I \uparrow$ → Light $\uparrow$

• Photodiode: photodiode converts light energy into electrical energy (it converts light into electrical current).

    Light energy → electrical energy

    Light $\uparrow$ → $I \uparrow$

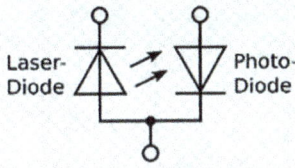

Figure 2.35   Light $\uparrow \rightarrow I \uparrow$

Table 2.3   *Photodiode vs. LED*

	Photodiode	**LED**
Symbol	Anode ▷┤ Cathode	▷┤
Energy	Light energy → electrical energy	Electrical energy → light energy
Biasing voltage	Reverse biased	Forward biased
Material	Si, Ge, InGaAs, GaP, etc.	GaAs, GaAsP, GaP, etc.
Function	It detects light	It emits light

2.2.4   Schottky diode

Introduction to Schottky diode

- Schottky diode (hot carrier diode): a diode formed by the junction of a semiconductor with a metal plate (a metal–semiconductor junction). It has a low forward voltage, fast switching time (high speed), and consumes less power.

    Walter H. **Schottky** (1886–1976): a German physicist who researched in solid-state physics and electronics and played a major early role in developing the theory of electron and ion emission phenomena.

- The symbol of the Schottky diode:

Figure 2.36   *Schottky diode*

- Construction: a semiconductor–metal junction is formed between an N-type semiconductor and a metal plate (metal replaces the P-type semiconductor).

    Schottky diode ≈ N-type semiconductor + metal plate

    - The metal side acts as the anode.
    - N-type semiconductor acts as the cathode.

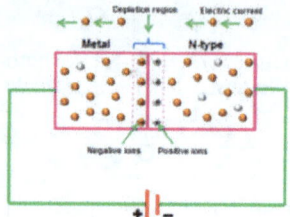

Figure 2.37   A Schottky diode circuit

Principle of operation

*   When a Schottky diode is forward biased, electrons in the N region cross the junction to the metal region. They join a large number of free electrons in the metal.
*   When a Schottky diode is reverse biased, the diode stops conducting almost immediately since a depletion region does not have to be established to block current flow.
*   Lower forward voltage: a lower forward bias drop in the Schottky diode is due to the metal. The resistance of the metal is less than the semiconductor (the voltage drop decreases with resistance).
*   Fast switching time: Schottky diode can be switched back and forth very quickly between forward and reverse operation since it has no depletion layer formed near the junction.
*   A lower forward voltage also allows higher switching speeds.
*   Majority carrier diode: Schottky diode operates with majority carriers (electrons – N type) and has no minority carriers (holes – P type).

Schottky diode vs. diode

Diode   Schottky diode

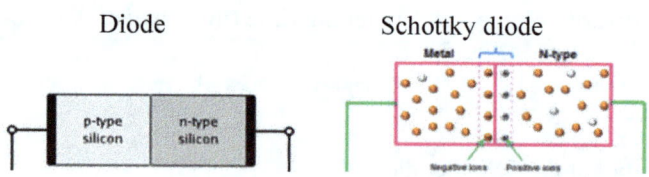

Figure 2.38   (a) Schottky diode vs. diode

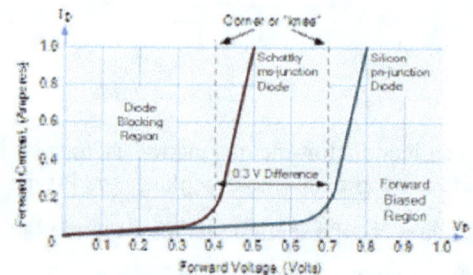

Figure 2.38 (b)   Schottky diode vs. diode – curve

Applications of Schottky diode

- Voltage clamping (fast clamping)
- Power rectifiers
- Solar cells
- Switched-mode power supplies
- Reverse current and discharge protection
- Sensitive communication receivers (such as radars)
- Radio frequency mixer and detectors
- Integrated circuits (TTL and CMOS logic gates)
- Digital computers, etc.

Summary

Overview of different types of diodes

- Laser diode (injection laser or diode laser): a diode that generates laser light of high intensity.
- BARITT diode (barrier injected transit-time diode): a high-frequency diode used for generating microwave signals. It uses thermionic emission rather than avalanche multiplication.
- Gunn diode: a high-frequency semiconductor component used for generating microwave signals. It has two terminals and formed by only N-type material.
- PIN diode: a special diode that has a layer of intrinsic semiconductor (no doping) that is sandwiched between a P-type and an N-type semiconductor material.
- Tunnel diode: a heavily doped diode that exhibits negative resistance, meaning the current decreases as the voltage increases. It works on the principle of the tunneling effect.

Zener diode

- Zener diode: a diode designed in such a way that it can work in the reverse breakdown region.
- In the Zener region: $I_R \nearrow \to \vec{V_z}$
- The Zener breakdown:

  $V_R \uparrow\uparrow \longrightarrow$ minority electrons accelerate $\longrightarrow$ collide with neighboring atoms $\longrightarrow$ knock valence electrons out of outermost shell $\longrightarrow$ numbers multiply $\longrightarrow$ avalanche $\longrightarrow I_R \uparrow\uparrow$

Zener voltage

- Zener voltage (V_z): the amount of reverse bias voltage that will make the P–N junction break down and allow the Zener diode to conduct in the reverse direction.
- Zener breakdown occurs with low Zener voltage V_z ratings ($V_z < 5$ V).
- Avalanche breakdown occurs at a higher value of Zener voltage V_z than does Zener breakdown ($V_z > 5$ V).

Breakdown characteristics

- The Zener knee current (I_{ZK}) or minimum current ($I_{Z(min)}$) is the turning point of the Zener diode into a breakdown. It is the minimum reverse current required to maintain constant breakdown voltage.
- Maximum Zener current (I_{ZM}): the maximum allowable amount of current that can pass through the Zener diode without damaging it.
- Zener test current (I_{ZT}): the current value to ensure good Zener voltage V_Z regulation.
- $I_{ZK} \leq I_Z \leq I_{ZM} \rightarrow V_Z \approx$ constant
- If $I_Z < I_{Zk} \rightarrow$ out of regulation

Equivalent circuit for Zener diode

- Equivalent circuit of a practical Zener diode: an actual Zener diode can be replaced by a voltage source in series with a resistor.
- Zener impedance: the equivalent series impedance of a practical Zener diode.

$$Z_Z = \frac{\Delta V_Z}{\Delta I_Z} = \frac{\Delta V_Z}{I_{ZT} - I_{Zk}}$$

Zener diode as a voltage shifter

- Voltage shifter (level shifter): a circuit used to convert voltage from one level to another.
- A Zener diode with a resistor can act as a voltage shifter. It has the ability to maintain a lower steady output voltage to the Zener diode's breakdown voltage.

Zener diode as a clipper (Zener limiter)

- A clipper is the circuit that can limit positive or negative (or both) amplitude of the input waveform to the desired output level.
- Three ways to limit waveform
  - Positive limiter: the positive peak of input sinusoidal voltage will be limited to the selected output Zener voltage.
  - Negative limiter: the negative peak of input sinusoidal voltage will be limited.
  - Back-to-back Zener limiter: with two opposing Zener diodes, the input sinusoidal waveform can be clipped to the Zener voltage on both polarities.

Zener diode as a voltage regulator

- Zener diodes are widely used as voltage regulators because of their reverse bias characteristics.
- In the reverse breakdown region, the voltage across the Zener diode (V_Z) is close to constant over a wide range of currents.

Line regulation

- Line regulation: the ability of a voltage regulator to maintain the output voltage V_{out} level with a varying input voltage V_{in}.

- $\overrightarrow{V_{in}} \rightarrow \overrightarrow{I} \rightarrow \overrightarrow{IR} \rightarrow \overrightarrow{V_Z} = \overrightarrow{V_{out}}$

Load regulation

- Load regulation: the ability of a voltage regulator to maintain the output voltage V_{out} level with a varying load.

- $\overrightarrow{R_L} \rightarrow \overrightarrow{I_L} \rightarrow \overrightarrow{I_Z} \rightarrow \overrightarrow{V_Z} = \overrightarrow{V_{out}}$

Percent regulation

- Percent regulation can be used to specify the performance of a voltage regulator. It can be in terms of line regulation or load regulation.

- Percent line regulation $= \left(\dfrac{\Delta V_{output}}{\Delta V_{input}} \right) 100\%$

- Percent load regulation $= \left(\dfrac{V_{No\ load} - V_{Full\ load}}{V_{Full\ load}} \right) 100\%$

Varactor diode

- Varactor diode (varactor diode): a specially designed diode whose junction capacitance varies with the variation of the reverse bias voltage (V_R).

    $\overrightarrow{V_R} \rightarrow \overrightarrow{C}$

- The varactor diode behaves as a capacitor. The N-type and the P-type regions can be considered to be the two metal plates.

- $\overrightarrow{V_R} \rightarrow$ depletion layer $\rightarrow$ capacitance

Light-emitting diode (LED)

- LED: a diode that emits light when current passes through it in the forward bias condition.
- LEDs produce different colors by using different semiconductor materials which produce photons at different wavelengths or frequencies.

Table 2.4   Semiconductor materials

Semiconductor materials	Produced light/heat
Si and Ge	Heat
GaAs	Infrared light (invisible light)
GaAsP	Red or yellow lights
GaP	Red, yellow, and green lights

Photodiode

- Photodiode (photodetector, photosensor, or light detector): a light-sensitive diode that converts light into electrical current. It is specially designed to operate in reverse bias conditions.
- Dark current I_D: a small reverse current that flows through a photodiode with no incident light.

Schottky diode

- Schottky diode (hot carrier diode): a diode formed by the junction of a semiconductor with a metal plate. It has a low forward voltage, fast switching time, and consumes less power.
- Construction: Schottky diode $\approx$ N-type semiconductor + metal plate
  - The metal side acts as the anode.
  - N-type semiconductor acts as the cathode.
- Majority carrier diode: Schottky diode operates with majority carriers and has no minority carriers.

Table 2.5   Schottky diode vs. diode

	Diode	**Schottky diode**
Symbol		
Junction	N-type and P-type semiconductor	Metal plate and N-type semiconductor
Forward voltage	Higher	Lower
Carrier	Majority and minority carriers (electrons and holes)	Majority carriers (electrons)
Switching speed	Slower	Faster
Function	It detects light	It emits light

Table 2.6   Types of diodes

Name	Symbol	Function	Biasing
Zener diode		It works in the reverse breakdown region	Reverse biasing
Varicap diode		The junction capacitance varies with the reverse bias voltage	Reverse biasing
LED		Emit light	Forward biasing
Photodiode		Detect light	Reverse biasing
Schottky diode		Emit light (It has a metal–N junction)	Forward biasing

Self-test

2.1 1. When the Zener voltage V_Z is exceeded and the Zener diode is operating in the Zener region, the voltage across the Zener diode stays (   ) even with the changing current.

2. The Zener breakdown occurs because of the heavy (   ) in the P–N junction.

3. Determine the Zener impedance of a practical Zener diode if the Zener voltage changes from 25 mV to 70 mV, and the Zener current changes from 7 mA (I_{ZK}) to 12 mA (I_{ZT}).

4. An actual Zener diode can be replaced by a voltage source in (   ) with a resistor.

5. A Zener diode has a resistor of 4 Ω. Determine the voltage across the Zener diode (Zener terminals) when the current is 16 mA and the voltage V_Z is 7 V.

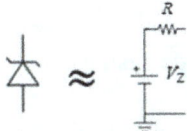

Figure 2.39   Ch 2: No. 5, self-test

6. A voltage (   ) is a circuit used to convert voltage from one level to another.

7. The (   ) is a circuit that can clip, limit, or cut off an input waveform to prevent the output waveform beyond a determined value.

8. The voltage (   ) is a circuit that is used to regulate the output voltage level into a lower, fixed level and remains constant for any changes in load or input voltage.

9. The back-to-back Zener limiter with two opposing Zener diodes and the input sinusoidal waveform can be clipped to the Zener voltage on (   ) polarities.

10. The (   ) regulation is the ability of a voltage regulator to maintain the output voltage V_{out} level with a varying input voltage V_{in}.

11. The voltage regulator in Figure 2.39 with a 4.6 V Zener diode at a Zener knee current (I_{ZK}) of 2 mA, Zener maximum current (I_{Zm}) of 152 mA, Zener test current (I_{ZT}) of 35 mA, Zener impedance of 4 Ω, and resistor R of 70 Ω. Determine the input voltages range (the minimum and maximum voltage) for the circuit that can be regulated.

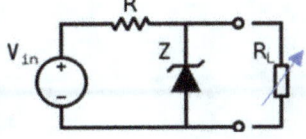

Figure 2.40   Ch 2: No. 11, self-test

12. The voltage regulator in Figure 2.41 with a 7 V Zener diode at a Zener knee current (I_{ZK}) of 1 mA. The resistor R is 80 Ω, and the supply voltage (V_{in}) is 18 V.
    a) Determine the maximum current for the circuit that can be regulated.
    b) Determine the minimum value of load resistance for the circuit that can be regulated.

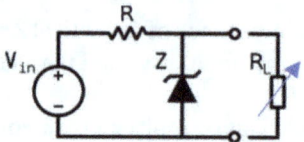

Figure 2.41   Ch 2: No. 12, self-test

13. Determine the percent voltage regulation if a voltage regulator with percent voltage with no load 15 V and percent voltage with full load 14 V.

2.2 14. A (   ) diode is a diode designed in such a way that it can work in the reverse breakdown region.

15. A varicap diode is a specially designed diode whose junction (   ) varies with the variation of the reverse bias voltage.

16. When a LED is (   ) biased, energy is released in the form of light.

17. The light-emitting semiconductor material determines the LED's (   ).

18. Gallium phosphide (GaP) emits red, yellow, and (   ) lights at different wavelengths.

19. A (   ) current is a small reverse current that flows through a photodiode with no incident light.

20. A (   ) diode is a light-sensitive diode that conducts current when a certain amount of light falls on it.

21. A photodiode converts light energy into (   ) energy.

22. A semiconductor–metal junction is formed between an N-type semiconductor and a (   ) plate.

23. A (   ) diode is a diode that has a very fast switching time and a low forward voltage drop.

Chapter 3

Diode applications – power supplies, clippers, and clampers

Chapter outline

3.1 Half-wave rectifiers

3.1.1 Introduction to DC power supply

DC power supply basics

- Direct current (DC) power supply: a power supply that can produce a constant output DC voltage to its load.
- Composition of a typical DC power supply: a basic DC power supply consists of four basic circuits, that is, transformer, rectifier, filter, and regulator as shown in Figure 3.1.

Figure 3.1   DC power supply

- Transformer: an electrical device formed by two coils that are wound on a common core. It can convert alternating current (AC) electrical energy from input to output.

  It can increase or decrease the output voltage or current (step-up or step-down).

- Rectifier: a device that converts AC signal to a pulsating DC signal.

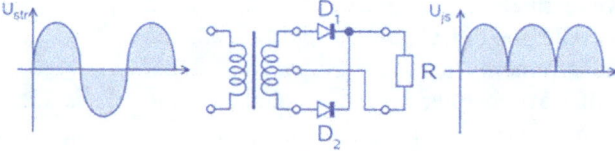

Figure 3.2   Rectifier

- Filter: a circuit that makes certain frequency components of a signal pass from the input and removes any unwanted frequency components.

  The output of the filter is a steady DC voltage, which has some ripples.

Figure 3.3   Filter waveforms

- Voltage regulator: a circuit that is designed to maintain the output voltage to a constant level despite changes in its input voltage or load.

  It can smooth the signal from the filter and produce a DC signal with no ripple.

3.1.2   Transformer

Introduction to transformer

- Transformer: a device formed by two coils (the primary and secondary) that are wound on a common core. It can transform AC electrical energy from input to output.

  It can increase or decrease the output voltage or current (AC).

Figure 3.4   Transformer

- Structurally, the transformers are categorized into two main types: the air-core and iron-core transformers. The symbols for them are shown in Figure 3.5(a) and (b), respectively.

Figure 3.5(a)   Air core   *Figure 3.5(b)   Iron core*

- Simplified transformer circuits: Figure 3.6 shows a simplified transformer circuit.
  - The primary winding is the coil connected to the AC power source.
  - The secondary winding is the coil connected to the load Z_L.

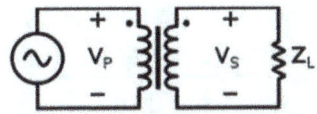

Figure 3.6   A transformer circuit

Power

- If the transformer is an ideal transformer, i.e., that transformer has no power loss itself, the input power (primary power P_{pri}) is equal to the output power (secondary power P_S).
- Calculating the power for an ideal transformer:

$$P_{pri} = P_{sec} \quad \text{or} \quad v_p i_p = v_s i_s, \quad \frac{i_P}{i_S} = \frac{v_S}{v_P} = n$$

  - v_p – is the primary voltage, v_s – is the secondary voltage, i_p – is the primary current, i_s – is the secondary current, and n – is the turns ratio.

Turns ratio

- Turns ratio (n): the ratio of turns of wire in the primary winding to the number of turns of wire in the secondary winding.

- Calculating the turns ratio (n): $n = \dfrac{N_s}{N_p} = \dfrac{v_S}{v_P} = \dfrac{i_P}{i_S}$

where N_p – is the Number of turns in the primary and N_S – is the Number of turns in the secondary.

Voltage and current

- The primary voltage (v_p): $$v_p = \frac{v_s}{n} \qquad n = \frac{v_s}{v_p}$$

- The primary current (i_p): $$i_p = ni_s \qquad \frac{i_p}{i_s} = n$$

Impedance

- The primary impedance (Z_p) can be obtained by substituting v_p and i_p into Z_p:

$$Z_P = \frac{v_P}{i_P} = \frac{v_S/n}{n\,i_S} = \frac{1}{n^2} Z_L \qquad Z = \frac{v}{i}$$

or $$n^2 = \frac{Z_L}{Z_P}, \qquad n = \sqrt{\frac{Z_L}{Z_P}}$$

- The secondary impedance is the load impedance Z_L: $Z_L = \dfrac{v_S}{i_S}$.

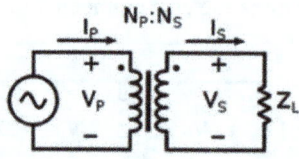

Figure 3.7   Currents and voltages in the transformer circuit

Step up, step-down, and center-tapped transformers

- A transformer is most commonly used to raise (step up) or reduce (step down) the output voltage of the circuits.

Figure 3.8a   Step-up transformer   Figure 3.8b   Step-down transformer

- A step-up transformer increases the output voltage. $V_{out} \uparrow$
  - $v_s > v_p$
  - $N_S > N_P$
  - $n > 1$
- A step-down transformer decreases the output voltage. $V_{out} \downarrow$
  - $v_s < v_p$
  - $N_S < N_P$
  - $n < 1$

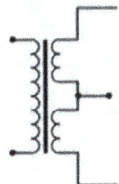

Figure 3.9 Center-tapped transformer

- A center-tapped transformer is designed to provide two separate and equal secondary voltages (a wire is connected across the exact middle point of the secondary winding of a transformer).

Transformer parameters conversion

- Conversion of the voltage, current, and impedance: a transformer can be used to convert voltage, current, and impedance.
- Voltage conversion:
  - Convert from the primary to the secondary, multiplying by n:

$$v_s = nv_p \qquad\qquad n = \frac{v_s}{v_p}$$

  - Convert from the secondary to the primary, multiplying by $\frac{1}{n}$:

$$v_p = \frac{1}{n}v_s \qquad\qquad n = \frac{v_s}{v_p}$$

- Current conversion:
  - Convert from the primary to the secondary, multiplying by $\frac{1}{n}$:

$$i_s = \frac{1}{n}i_p \qquad\qquad n = \frac{i_p}{i_s}$$

  - Convert from the secondary to the primary, multiplying by n:

$$i_p = n\, i_s \qquad\qquad n = \frac{i_p}{i_s}$$

- Impedance conversion:
  - Convert from the primary to the secondary, multiplying by $\frac{1}{n^2}$:

$$Z_P = \frac{1}{n^2}Z_L \qquad\qquad Z_L = n^2 Z_P$$

  - Convert from the secondary to the primary, multiplying by n^2:

$$Z_L = n^2 Z_P$$

Table 3.1   Transformer parameters conversion

Parameters conversion	Formulas
Voltage conversion	$v_s = n v_p, \; v_p = \dfrac{1}{n} v_s$
Current conversion	$i_s = \dfrac{1}{n} i_p, \; i_p = n i_s$
Impedance conversion	$Z_p = \dfrac{1}{n^2} Z_L, \; Z_L = n^2 Z_p$

The main applications of transformers
- Increase or decrease the voltage or current.
- Transfer electric energy from one circuit to another.
- Prevent DC from passing from one circuit to the other.
- Isolate two circuits electrically.
- Impedance matching.
- … …

3.1.3   Half-wave rectifier
Introduction to half-wave rectifier
- Rectifier: a device that converts AC signal to a pulsating DC signal.

Figure 3.10   AC to pulsating DC

- Half-wave rectifier: a type of rectifier circuit that allows one half-cycle of an AC signal to pass. It can remove the negative component of an alternating signal.
- A basic half-wave rectifier circuit (the diode allows the current to flow only in one direction).

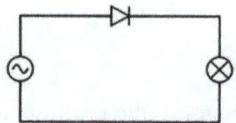

Figure 3.11   A half-wave rectifier circuit

Operation of a half-wave rectifier

- Positive half-cycle: the diode is forward biased (FB) and is conducting. The input signal is produced at the output (the output will be the same as the positive half-cycle of the input).
  - During the positive half-cycle, the input voltage must overcome the barrier potential V_{BR} before the diode becomes FB.
  - The peak value of the output voltage ($V_{pk(out)}$) is 0.7 V (Si) or 03 V (Ge.) less than the peak value of the input ($V_{pk(in)}$).
  - The "on" state of the diode: $V_{pk(out)} = V_{pk(in)} - V_{BR} \approx V_{pk(in)}$

    $$V_{BR} = 0.7V \text{ (or } 0.3V) \approx 0$$
  - Waveform:

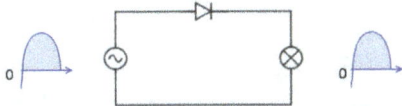

Figure 3.12   Positive half cycle

- Negative half-cycle: the diode is reverse biased (RB), which blocks the input signal (no current).
  - The "off" state of diode: $I \approx 0$, $V_{out} = I R_L = 0$

    The RB diode current is very small and very close to zero.
  - Waveform:

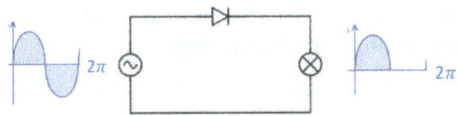

Figure 3.13   Negative half-cycle ($V_{out} = 0$)

- Output voltage: the output voltage is only the positive half-cycles of the AC input voltage (a pulsating DC voltage). The cycle repeats.

Figure 3.14   Output voltage

The average output voltage of a half-wave rectifier

- The average output voltage (this voltage could be measured on a multimeter): it is determined by finding the area under the curve over a full cycle from 0 to 2π ($V_{out} = 0$ for the negative cycle).

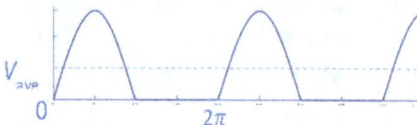

Figure 3.15   Average output voltage

- Calculating the average output voltage:

$$V_{avg} = \frac{V_{pk(out)}}{\pi} \approx 0.318 \, V_{pk(out)} \qquad V_{pk(out)} - \text{the peak output voltage}$$

Example: $V_{pk(out)} = 120 \text{ V}, V_{avg} = ?$

$$V_{avg} = \frac{V_{pk\,(out)}}{\pi} = \frac{120 \text{ V}}{\pi} \approx 38.2 \text{ V}$$

Or $V_{avg} = 0.318 \, V_{pk(out)} = (0.318)\,(120 \text{ V}) \approx 38.2 \text{ V}$

The peak inverse voltage of a half-wave rectifier

- Peak inverse voltage (PIV): the maximum voltage a diode can withstand in the RB direction (the maximum reverse voltage).

  The PIV occurs at the peak of the input cycle when the diode is RB.

- Calculating the PIV: $PIV = V_{pk\,(in)}$

  $V_{pk\,(in)}$ – the peak input voltage (for negative or reverse cycle)

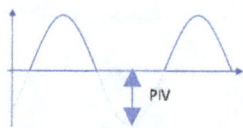

Figure 3.16   Peak inverse voltage

A transformer-coupled half-wave rectifier

- The transformer-coupled half-wave rectifier uses a step-down transformer to step down the input AC signal (normally AC input voltage will be high). The AC source is also electrically isolated from the rectifier.

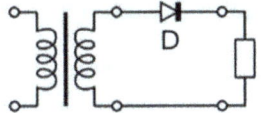

Figure 3.17   A transformer-coupled half-wave rectifier

- The secondary voltage (V_{sec}) of the transformer:

$$V_{sec} = \frac{N_{Sec}}{N_{Pri}} V_{pri} = n\, V_{pri} \qquad\qquad n = \frac{N_s}{N_p} = \frac{v_s}{v_p}$$

- The pick output voltage ($V_{pk(out)}$) of the half-wave rectifier:
  - With transformer: $V_{pk(out)} = V_{pk(sec)} - V_{BR}$

  $\qquad\qquad V_{BR}$ – barrier potential ($V_{BR} = 0.7$ V for Si or 0.3 V for Ge)

  - Without transformer: $V_{pk(out)} = V_{pk(in)} - V_{BR}$
- Calculating the PIV: $\text{PIV} = V_{pk\,(sec)}$

Example: Determine the peak value of the output voltage and the PIV for Figure 3.17, if the turns ratio of the transformer is 0.33 and the peak input voltage is 120 V.

(A silicon diode)

- Given: $\qquad n = \dfrac{N_{Sec}}{N_{Pri}} = \ 0.33,\ V_{pk(in)} = 120$ V

- Find: $\qquad V_{pk(out)}$ and PIV

- Solution: $\qquad n = \dfrac{N_{Sec}}{N_{Pri}} = 0.33 \approx \dfrac{1}{3}$

$$V_{pk(sec)} = \frac{N_{sec}}{N_{Pri}} V_{pk(pri)} = \frac{1}{3}(120\text{V}) = 40 \text{ V} \qquad V_{Pk(pri)} = V_{pk(in)}$$

$$V_{pk(out)} = V_{pk(sec)} - V_{BR} = 40\text{V} - 0.7 \text{ V} = 39.3 \text{ V}$$

$$\text{PIV} = V_{pk(sec)} = 40 \text{ V}$$

3.2   Full-wave rectifier

3.2.1   A center-tapped full-wave rectifier

Full-wave rectifier

- Full-wave rectifier: a type of rectifier circuit that converts the entire cycle of the AC signal into pulsating DC (using both positive and negative half-cycles of the applied sine wave).

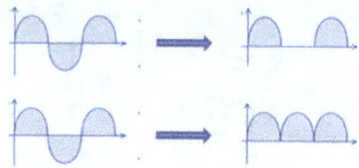

Figure 3.18   Halfwave vs. full wave

- A center-tapped full-wave rectifier circuit: two diodes + a center-tapped transformer

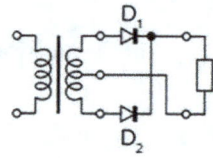

Figure 3.19   A full wave rectifier

    – Two diodes: one conducts during one half-cycle while the other conducts during the other half-cycle of the input AC voltage.
    – A center-tapped transformer: provides two separate and equal secondary voltages for two diodes.

Operation of a center-tapped full-wave rectifier

- Positive half-cycle: the diode D_1 is FB and can conduct. The diode D_2 is RB and cannot conduct.

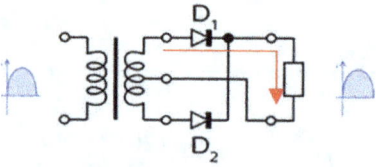

Figure 3.20(a)   Positive half-cycle

    – D_1: FB – the diode D_1 is on
    – D_2: RB – the diode D_2 is off
- Negative half-cycle: the diode D_2 is FB and can conduct.

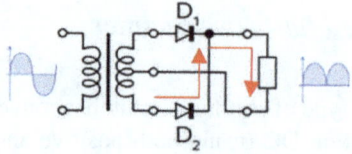

Figure 3.20(b)   Negative half-cycle

The diode D_1 is RB and cannot conduct.

- D_2: FB – the diode D_2 is on
- D_1: RB – the diode D_1 is off

• The cycle repeats.

Figure 3.21 Full wave

Effect of the turns ratio on the output voltage

• If the transformer's turn's ratio n is 1 (1:1), the output voltage (V_{out}) approximately equals half the primary input voltage (V_{Pri}).

Half of the primary voltage appears across each half of the secondary winding.

– The secondary voltage: $V_{sec} = n\, V_{pri} = (1)\,(V_{pri}) = V_{pri}, \; V_{sec} = V_{pri}$ $n = \dfrac{V_{sec}}{V_{pri}}$

– The pick output voltage: $V_{pk(out)} \approx \dfrac{1}{2} V_{pk(pri)} \approx \dfrac{1}{2} V_{pk(sec)}$

$$V_{pk(out)} = \dfrac{1}{2} V_{pk(sec)} - V_{BR} \qquad V_{BR} = 0.7\text{V (Si) or } 0.3\text{V (Ge)}$$

• To obtain an output voltage equal to the input ($V_{out} = V_{in}$): use a 1:2 step-up center-tapped transformer. If the transformer's turn's ratio n is 2 (1:2), the total secondary voltage (V_{Sec}) is twice the primary voltage ($2V_{Pri}$).

– The secondary voltage: $n = \dfrac{V_{sec}}{V_{pri}}, \; V_{sec} = n\, V_{pri} = 2\, V_{pri}, \; V_{pri} = \dfrac{1}{2} V_{sec}$

$$V_{sec} = 2\, V_{pri}$$

– The peak output voltage: $V_{pk(out)} \approx \dfrac{1}{2} V_{pk(sec)} \approx \dfrac{1}{2} 2\, V_{pk(pri)} \approx V_{pk(pri)}$

$$V_{pk(out)} = \dfrac{1}{2} V_{pk(sec)} - V_{BR} \qquad V_{BR} = 0.7\text{ V (Si) or } 0.3\text{ V (Ge)}$$

• General case (no matter what the turn's ratio): the output voltage of a center-tapped full-wave rectifier is always one-half of the total secondary voltage less the diode drop.

The output voltage: $V_{pk(out)} \approx \dfrac{1}{2} V_{pk(sec)}$

$$V_{pk(out)} = \dfrac{1}{2} V_{pk(sec)} - V_{BR} \qquad V_{BR} = 0.7\text{ V (Si) or } 0.3\text{ V (Ge)}$$

Peak inverse voltage

- Peak inverse voltage (PIV) of a full-wave rectifier: the maximum reverse voltage that each diode must withstand for a full-wave rectifier is the peak secondary voltage ($V_{(pk\,sec)}$) the less the diode drop.
- Calculating the PIV:
  - Half-wave: $PIV = V_{pk(in)}$: without transformer

    $$PIV = V_{pk(sec)} : \text{with transformer}$$

  - Full-wave: $PIV = V_{pk(sec)} - V_{BR}$ (with transformer)

$$V_{BR} = 0.7V \text{ (Si) or } 0.3V \text{ (Ge)}$$

$$KVL: \Sigma V = 0, 0.7\,V + PIV - V_{P(sec)} = 0$$

Or $\qquad\qquad PIV = 2\,V_{Pk(out)} + V_{BR}$ $\qquad\qquad$ PIV in terms of $V_{Pk(out)}$

Derive: $\qquad V_{Pk(out)} = \dfrac{1}{2}V_{pk(sec)} - V_{BR}$

$$V_{pk(out)} = \frac{1}{2}\left(PIV + V_{BR}\right) - V_{BR} \qquad V_{pk(sec)} = PIV + V_{BR}\ (PIV = V_{pk(sec)} - V_{BR})$$

$$2V_{pk(out)} = PIV + V_{BR} - 2\,V_{BR} \qquad\qquad \text{Multiply by 2}$$

$$2V_{pk(out)} = PIV - V_{BR}$$

$$PIV = 2\,V_{pk(out)} + V_{BR}$$

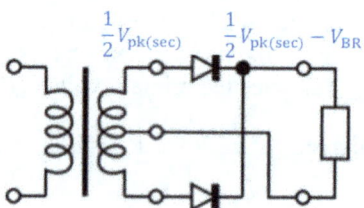

Figure 3.22   PIV

Calculating the average output voltage (V_{AVG}):

- Half-wave: $V_{AVG} = \dfrac{V_{pk(out)}}{\pi} \approx 0.318\,V_{pk(out)}$

- Full-wave: $V_{AVG} = \dfrac{2\,V_{pk(out)}}{\pi} \approx 0.637\,V_{pk(out)}$ $\qquad\qquad 2\,V_{Pk(out)} - \text{two half-wave}$

Example: Determine the peak secondary voltage, the peak output voltage, the PIV, and the average output voltage for the circuit in Figure 3.21, if the turns ratio of the transformer is 2 (1:2) and the peak input voltage is 110 V. (Assuming two silicon diodes.)

- Given: $n = 2$, $V_{\text{pk (pri)}} = 110$ V
- Find: $V_{\text{pk(sec)}}$, $V_{\text{pk(out)}}$, PIV, and V_{AVG}.
- Solution:

$V_{\text{pk(sec)}}$: $\quad V_{\text{pk(sec)}} = \dfrac{N_S}{N_P} V_{\text{pk(pri)}} = \dfrac{2}{1}(110 \text{ V}) = 220 \text{ V}$ $\qquad n = \dfrac{N_s}{N_p} = 2$

$V_{\text{pk(out)}}$: $\quad V_{\text{pk(out)}} = \dfrac{1}{2} V_{\text{pk(sec)}} - V_{\text{BR}} = \dfrac{1}{2}(220\text{V}) - 0.7 \text{ V} = 109.3 \text{ V}$

PIV: $\quad \text{PIV} = V_{\text{pk(sec)}} - V_{\text{BR}} = 220 \text{ V} - 0.7 \text{ V} = 219.3 \text{ V}$

V_{AVG}: $\quad V_{\text{AVG}} = \dfrac{2\, V_{\text{pk(out)}}}{\pi} \approx 0.637\, V_{\text{pk(out)}} \approx (0.637)(109.3 \text{ V}) \approx 69.6 \text{ V}$

3.2.2 Full-wave bridge rectifier

Introduction to full-wave bridge rectifier

- Rectifiers:
  - A half-wave rectifier uses one diode.
  - A full-wave rectifier uses two diodes.
  - A full-wave bridge rectifier uses four diodes.

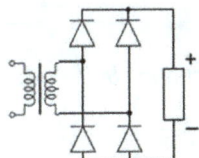

Figure 3.23 Full-wave bridge rectifier

- Bridge rectifier: a type of full-wave rectifier circuit that uses four diodes connected in a closed-loop "bridge" configuration to efficiently convert entire cycle of the AC signal into the DC.
- A bridge rectifier circuit: four diodes + a center-tapped transformer

Operation of the bridge rectifier

- Operation: two diodes conducting current during each half-cycle.
- Positive half-cycle: diodes D_1 and D_4 are FB and can conduct current. Diodes D_2 and D_3 are RB and not conducting.
  - D_1 and D_4: FB – the diodes D_1 and D_4 are on
  - D_2 and D_3: RB – the diodes D_2 and D_3 are off

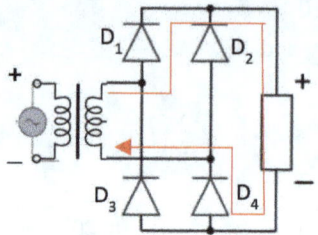

Figure 3.24   Positive half-cycle – D_1 and D_4 on

- Negative half-cycle: diodes D_2 and D_3 are FB and can conduct current. Diodes D_1 and D_4 are RB and not conducting.
  - D_2 and D_3: FB – the diodes D_2 and D_3 are on
  - D_1 and D_4: RB – the diodes D_1 and D_4 are off

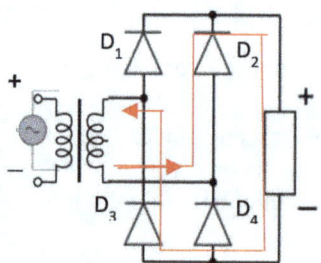

Figure 3.25   Negative half-cycle – D_2 and D_3 on

The output voltage of a bridge rectifier
- Neglect diode drops: $V_{pk\,(out)} \approx V_{pk(sec)}$
- With the diode drops: $V_{pk\,(out)} = V_{pk(sec)} - 2V_{BR}$ $V_{BR} = 0.7$ V (Si) or 0.3 V (Ge)

  Derive: $V_{BR} + V_{out} + V_{BR} = 0$ KVL: $\Sigma V = 0$

PIV of a bridge rectifier
- Neglect diode drops: PIV $\approx V_{pk\,(out)}$ The same as a center-tapped full-wave rectifier.
- With the diode drops: PIV $= V_{pk\,(out)} + V_{BR}$ $V_{BR} = 0.7$ V (Si) or 0.3 V (Ge)

      Comparing with a center-taped full-wave rectifier.

The main advantage of the bridge rectifier
- An ordinary transformer is used in place of a center-tapped transformer (reducing size and cost).
- The output is almost twice that of the center-tapped full-wave rectifier.

- Full-wave center-taped rectifier: $V_{pk(out)} = \dfrac{1}{2}V_{pk(sec)} - V_{BR}$

$$2V_{pk(out)} = V_{pk(sec)} - 2_{VBR}$$

- Bridge rectifier: $V_{pk(out)} = V_{pk(sec)} - 2\,V_{BR}$

• The PIV is one-half that of the full-wave center-tapped rectifier.
  - Full-wave center-tapped rectifier: $PIV = 2V_{Pk(out)} + V_{BR}$
  - Bridge rectifier: $PIV = V_{pk\,(out)} + V_{BR}$

Example: Determine the peak output voltage and the PIV for the bridge rectifier circuit in Figure 3.26, if the turns ratio of the transformer is 2 (1:2) and the peak input voltage is 110 V. (Four silicon diodes)

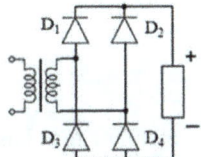

Figure 3.26   Full-wave bridge rectifier – an example

- Given: A 1:2 transformer, $V_{pk\,(pri)}$ = 110 V
- Find: $V_{pk(out)}$ and PIV
- Solution:

$$V_{pk(out)} : V_{pk(sec)} = \frac{N_S}{N_P}V_{pk(pri)} = \frac{2}{1}\,(110\ \text{V}) = 220\ \text{V}$$

$$V_{pk(out)} = V_{pk(sec)} - 2\,V_{BR} \qquad\qquad V_{BR} = 0.7\ \text{V (Si)}$$

$$= 220\ \text{V} - 1.4\text{V} = 218.6\ \text{V}$$

PIV: $PIV = V_{pk\,(out)} + V_{BR}$

$$= 218.6\ \text{V} + 0.7\ \text{V} = 219.3\ \text{V}$$

3.3   Power supply filters

3.3.1   Capacitor filter

Filter

• Recall – composition of a typical DC power supply: a basic DC power supply consists of four basic circuits, i.e. transformer, rectifier, filter, and regulator.

• Filter: a circuit that makes certain frequency components of a signal pass from the input signal and removes any unwanted frequency components.

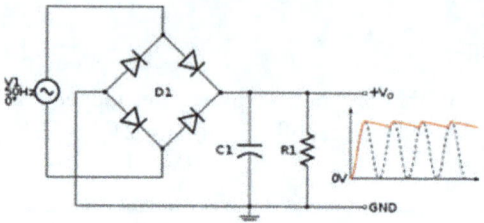

Figure 3.27   Filter

• In power supplies, filters are used to smooth out the pulsating DC output obtained from the rectifier circuit.

Capacitor

• Capacitor: an energy storage element that has two parallel conductive metal plates separated by an isolating material (the dielectric). It can store and release charges that it absorbed from the power supply.
• Charging and discharging a capacitor: once the three-position switch is turned on to position 1, current I will flow in the circuit – charging. When the switch is closed to position 2, the capacitor releases its charges – discharging.
  − Charging: the process of storing energy (the switch is turned on to position 1).

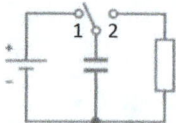

Figure 3.28(a)   Charging a capacitor

  − Discharging: the process of releasing energy (the switch is closed to position 2).

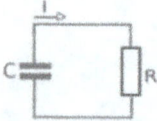

Figure 3.28(b)   Discharging a capacitor

3.3.2 Operation of the capacitor filter

Half-wave rectifier with capacitor filter

- The half-wave rectifier converts the AC signal into a pulsating DC with AC ripples.
- A capacitor filter (a capacitor connected in parallel with the output of the rectifier) can be used to reduce the AC components and make the output of rectifier smooth at the output.

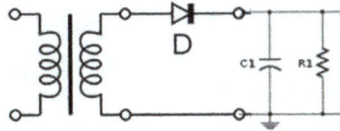

Figure 3.29 Half-wave rectifier with capacitor filter

Operation of the capacitor filter

- During the period of t_1-t_2: the capacitor charges from 0 V to $V_{pk(in)} - V_{BR}$ (when the diode is on).
- During the period of t_2-t_3: the capacitor discharges (when the diode is off).

Figure 3.30 Output voltage

RC time constant τ

- In an *RC* circuit, the charging and discharging is a gradual process that needs some time.
- The time rate of this process depends on the values of circuit capacitance *C* and resistance *R*. The variation of the *R* and *C* will affect the rate of charging and discharging.
- The product of the *R* and *C* is called the *RC* time constant, and it can be expressed as a Greek letter τ (tau), i.e. τ = *RC*.
- The time constant is the time interval required for a system or circuit to change from one state to another, i.e. the time required to charge or discharge in an *RC* circuit.

Calculating *RC* time constant τ

- *RC* time constant τ = (Resistance) (Capacitance)   or   τ = *RC*
- Units: Second (s)
                    → τ = *RC* ←──────── Farad (F)
          Ohm (Ω)

Ripple voltage

- Ripple voltage: a small, periodic variation in a DC voltage within a power supply that remains after rectification and filtering of an AC voltage.
- Larger and smaller ripples:

Figure 3.31   Larger and smaller ripples

RC time constant and ripple voltage

- High *RC* time constant τ leads to a longer charging or discharging time.
- High *RC* time constant τ leads to a lower ripple of the output voltage.

$$\tau\uparrow \; (R\uparrow \text{ or } C\uparrow) \rightarrow \text{ripple voltage} \downarrow$$

- Low RC time constant τ leads to a larger ripple of the output voltage.

$$\tau\downarrow \; (R\downarrow \text{ or } C\downarrow) \rightarrow \text{ripple voltage} \uparrow$$

Output frequency

- Half- and full-wave ripple: a filtered full-wave rectifier voltage has much less ripple than a half-wave rectifier voltage.
  - Half-wave ripple:

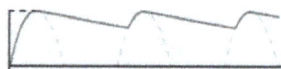

Figure 3.32(a)   Half-wave ripple

  - Full-wave ripple:

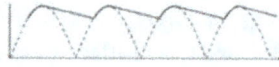

Figure 3.32(b)   Full-wave ripple

- Frequency: the output frequency of a full-wave rectifier is twice that of a half-wave rectifier.
- Period (*T*): $T_{\text{Half}} = 2\,T_{\text{Full}}$

- Frequency (*f*): $f = \dfrac{1}{T}, \quad f_{\text{Full}} = \dfrac{1}{T_{\text{Full}}} = \dfrac{1}{\frac{1}{2}T_{\text{Half}}} = 2\,f_{\text{Half}}$

3.3.3 Ripple factor and surge current

Ripple factor

- Ripple factor (RF): the ratio of the AC component's root mean square (RMS) value or peak-to-peak value to the DC component of the output voltage.

$$\text{Ripple factor} = \frac{\text{The AC component of the output} \left(\text{RMS value}\right)}{\text{The DC component of the output}} \quad \text{or}$$

$$\text{RF} = \frac{V_{AC}}{V_{DC}} = \frac{I_{AC}}{I_{DC}}$$

- The lower the RF (the lesser the AC ripple), the better the filter.
  RF$\downarrow$→ better filter
- To reduce the RF: increase the time constant (the output can be made smoother).

$$\tau\uparrow (R\uparrow\text{or } C\uparrow) \rightarrow \text{RF} \downarrow \rightarrow \text{ripple voltage} \downarrow$$

Surge current

- Surge current (inrush current or switch-on surge): the instantaneous rise current in a short period of time in a power supply (or electrical equipment) at turn-on.
- Filter capacitor and surge current: when a power supply is initially turned on, charging filter capacitor can result in a surge current.
- Use surge-limiting resistor: a resistor in series with the diode can be used to limit surge current.
  - to avoid blowing fuses or tripping circuit breakers.
  - to avoid damage diode.
- $\text{Minimum surge resistor value} = \dfrac{\text{Peak secondary voltage} - 2\,V_{BR}}{\text{Diode max forward surge current} \left(\text{from data sheet}\right)}$

$$\text{or} \quad R_{(surge)} = \frac{V_{pk(sec)} - 2\,V_{BR}}{I_{FSM}}$$

3.4 Diode clipping and clamping circuits

3.4.1 Clippers

Diode clipping circuit

- Clipper (limiter): a circuit that limits or clips off a portion of an input AC signal above or below certain levels.
- Positive clipper: a clipper circuit that removes the positive half-cycles of the output waveform.

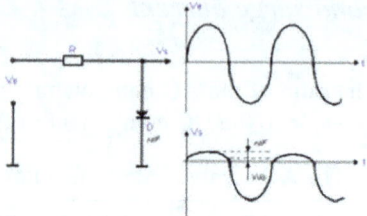

Figure 3.33   Positive clipper

— During the positive half-cycle: the diode D is FB and conducts current (it is on). The output is limited to V_{BR}.

> The diode D: FB, D is on $V_{out} = V_{BR}$ $V_{BR} = 0.7$ V (Si) or 0.3 V (Ge)

— During the negative half-cycle: the diode D is RB and is not conducting (it is off). The output is approximate to input.

> The diode D: RB, D is off, $V_{out} = V_{in}$

• Negative clipper: a clipper circuit that removes the negative half-cycles of the output waveform.   The diode is turned around.

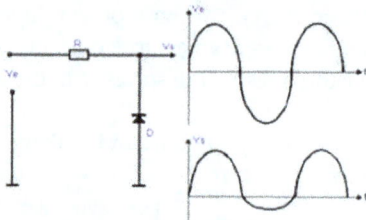

Figure 3.34   Negative clipper

— During the positive half-cycle: the diode D is RB and is not conducting (it is off). The output is approximate to input.

> The diode D: RB, D is off, $V_{out} = V_{in}$

— During the negative half-cycle: the diode D is FB and conducts current (it is on). The output is limited to $-V_{BR}$.

> The diode D: FB, D is on $V_{out} = -V_{BR}$

The output voltage of clippers

• The output voltage looks like the input voltage, but with a magnitude determined by the voltage divider formed by R and the load resistor R_L.
• Calculating the output voltage: Voltage divider rule

$$V_{out} = V_{in} \frac{R_L}{R_L + R}$$

If $R \ll R_L$, then $V_{out} = V_{in}$

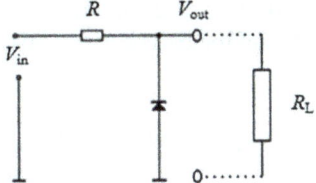

Figure 3.35   Negative clipper with a load

Example: A negative clipper shown in Figure 3.35 has an RMS input voltage of 8.5 V, R is 5 kΩ, and the load resistor is 50 kΩ. What is the peak output voltage for the clipper? (Assuming a silicon diode)

- Given: $V_{rms\ (in)} = 8.5$ V, $R = 5$ kΩ, and $R_L = 50$ kΩ.
- Find: $V_{pk\ (out)} = ?$
- Solution:

$V_{pk(in)}$: $V_{pk(in)} = \sqrt{2}\ V_{rms(in)} = \sqrt{2}\ (8.5\ \text{V}) \approx 12\ \text{V}$ $V_{pk} = \sqrt{2}\ V_{rms}$

$V_{pk\ (out)}$: $V_{pk\ (out)} = V_{pk\ (in)}\ \dfrac{R_L}{R_L + R} = 12\ \text{V}\ \dfrac{50\ \text{kΩ}}{50\ \text{kΩ} + 5\ \text{kΩ}} \approx \mathbf{10.91\ V}$

Positive half-cycle: Vpk (out) ≈ 10.91 V
Negative half-cycle: $V_{pk\ (out)} = -V_{BR} = -0.7$ V

Biased clippers

- Biased clipper: a clipper circuit that uses a DC voltage source to adjust the limiting level of the output voltage.
- It biases the diode and removes a small portion of positive or negative half cycles of the input voltage.
- The value of the DC voltage source determines the voltage at which the diode begins conducting.

Biased positive and negative clippers

- Biased positive clipper: a clipper circuit that adjusts the positive half-cycles of the limiting level.

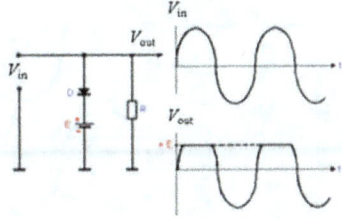

Figure 3.36   Biased positive clipper

   – During the positive half-cycle, the sum of the positive-biased voltage
     E (DC source voltage) and the barrier potential V_{BR} will be the output.

$$V_{out} = E + V_{BR}$$ The diode is on.

   – During the negative half-cycle: the output is approximate to input.

$$V_{out} = V_{in}$$ The diode is off.

- Biased negative clipper: a clipper circuit that adjusts the negative half-cycles of
  the limiting level.

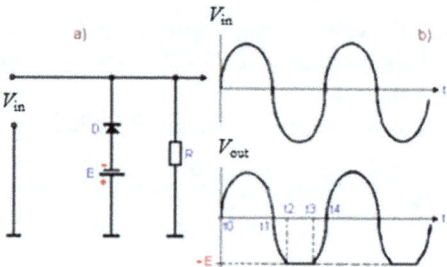

Figure 3.37   Biased negative clipper

   – During the positive half-cycle, the output is approximate to input.

$$V_{out} = V_{in}$$ The diode is off.

   – During the negative half-cycle, the sum of a negative-biased voltage
     ($-E$, DC voltage source voltage) and the negative barrier potential will
     be the output.

$$V_{out} = -E - V_{BR}$$ The diode is on.

Combination clipper

- Combination clipper: a clipper circuit that adjusts both the positive and negative half-cycles of the limiting level.

  Both half-cycles can be adjusted using a single circuit.

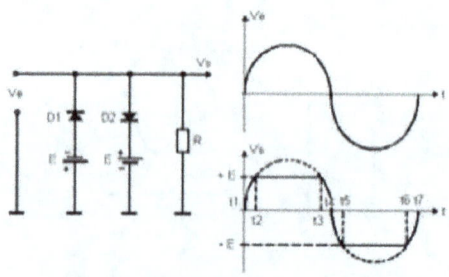

Figure 3.38   Combination clipper

- During the positive half-cycle, the sum of the positive-biased voltage E (DC source voltage) and the barrier potential will be the output.

$$V_{out} = E + V_{BR}$$ D1 is on.

- During the negative half-cycle, the sum of a negative-biased voltage $(-E)$ and the negative barrier potential will be the output.

$$V_{out} = -V_{BS} - V_{BR}$$ D2 is on.

Applications of clippers
- Amplitude selectors
- In power supplies
- In FM transmitters (for removing the excess ripples in the signals)
- To separate synchronizing signals from the composite picture signals
- To clip the excessive noise spikes
- To protect the transistor from transients
- To generate new waveforms or shape the existing waveform, etc.

3.4.2   Clampers

Diode clamping circuit
- Clamper: a circuit that adds a DC level to an AC signal and moves the whole signal up or down.
- Clamper vs. clipper   They are opposite to each other regarding their working principle.
  - Clipper: used to remove (limit) a portion of an AC signal. (The voltage that is clipped by clipper can change in shape – an AC signal + a DC level.)
  - Clamper: used to add (shift) a DC level to an input AC signal. (The output voltage obtained by clamper does not change in shape – it moves the whole signal up or down.)

Positive clamper
- A positive clamper is a circuit that shifts an entire input waveform above (a negative peak of the signal is shifted above the x-axis).

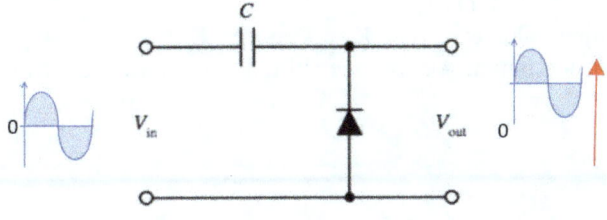

Figure 3.39   Positive clamper

- During the negative half-cycle, the diode D is FB. The capacitor is charged to input voltage $V_{pk(in)}$. The voltage across the capacitor is the sum of the input voltage and the negative V_{BR}.
  - The diode D: FB, D is on, C charges.
  - The capacitor's voltage: $V_C = V_{pk(in)} - V_{BR}$
- During the positive half-cycle, the diode D is RB. The output voltage is the sum of the input voltage and the capacitor's voltage.
  - The diode D: RB, D is off.
  - The output voltage: $V_{pk(out)} = V_C + V_{pk(in)}$

    The capacitor holds its changed voltage V_C as a battery in series with an input voltage. The load resistor R_L must be sufficiently high ($\tau = R_L C$) to prevent significantly capacitor discharging.

- As a result, the output AC voltage shifted upward – shift the input signal $(V_C + V_{pk(in)})$.

Negative clamper

- A negative clamper is a circuit that shifts an entire input waveform below a DC voltage (a positive peak of the signal is shifted below the *x*-axis).

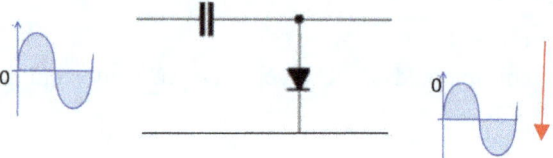

Figure 3.40   Negative clamper

- During the positive half-cycle, the diode D is FB (it is on). The capacitor's voltage is charged to the sum of the input voltage and the negative barrier potential.
  - The diode D: FB, D is on, C charges.
  - The capacitor's voltage: $V_C = V_{pk(in)} - V_{BR}$

    $(V_C \neq V_{pk(in)})$

- During the negative half-cycle, the diode D is RB (it is off). The output voltage is the sum of the input voltage and the negative capacitor voltage.
  - The diode D: RB, D is off,
  - The output voltage: $V_{pk(out)} = -V_C + V_{pk(in)}$
- As a result, the output AC voltage shifted downward – shift the input signal $(-V_C + V_{pk(in)})$.

Example: A negative clamper shown in Figure 3.40 has a peak negative input voltage of 12 V. What is the RMS output voltage for the clamper? (Assuming a silicon diode)

- Given: $V_{pk\,(in)} = -12$ V
- Find: $V_{RMS\,(out)}$
- Solution:
  V_C:
  $$V_C = V_{pk(in)} - V_{BR} = V_{pk(in)} - 0.7 \text{ V}$$
  $$= -12 \text{ V} - 0.7 \text{ V} = -12.7 \text{ V}$$

  $V_{pk(out)}$:
  $$V_{pk(out)} = -V_C + V_{pk(in)} = -12.7 \text{ V} + (-12 \text{ V}) = -24.7 \text{ V}$$

  $V_{RMS(out)}$:
  $$V_{RMS(out)} = \frac{V_{pk\,(out)}}{\sqrt{2}} = \frac{-24.7 \text{ V}}{\sqrt{2}} \approx \mathbf{-17.5 \text{ V}} \qquad V_{pk} = \sqrt{2}\, V_{rms}$$

Biased clampers

- Biased clamper: a clamper circuit that uses a DC voltage source to have an additional shift level of the output voltage.
- The working principle of the biased clampers is similar to the biased clippers (a DC voltage source is added to a biased clamper).

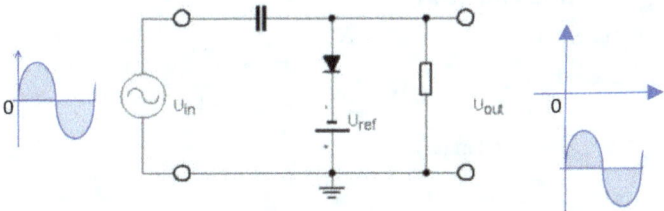

Figure 3.41   Biased clamper

Applications of campers

- Voltage doublers or voltage multipliers
- Test equipment
- Sonar systems
- Baseline stabilizer (in transmitter and receiver)
- To improve reverse recovery time
- To remove distortion
- To protect amplifier (from large errant signals)
- Etc.

Summary

DC power supply basics
- DC power supply: a power supply that can produce a constant output DC voltage to its load.
- Composition of a typical DC power supply: a basic DC power supply consists of four basic circuits, i.e. transformer, rectifier, filter, and regulator.
- Transformer: an electrical device formed by two coils that are wound on a common core. It can convert AC electrical energy from input to output.
- Rectifier: a device that converts AC signal to DC signal (a pulsating DC).
- Filter: a circuit that makes certain frequency components of a signal pass from the input and removes any unwanted frequency components.
- Voltage regulator: a circuit that is designed to maintain the output voltage to a constant level despite changes in its input voltage or load.

Power supply filters
- In power supplies, filters are used to smooth out the pulsating DC output obtained from the rectifier circuit.
- A capacitor filter can be used to reduce the AC components and make the output of rectifier smooth at the output.

Calculating *RC* time constant τ
- *RC* time constant: $\qquad\qquad \tau = RC$

- Units:

$$\text{Second (s)} \searrow$$
$$\tau = RC \longrightarrow \text{Farad (F)}$$
$$\text{Ohm } (\Omega) \nearrow$$

Ripple voltage
- Ripple voltage: a small, periodic variation in a DC voltage within a power supply that remains after rectification and filtering of an AC voltage.
- *RC* time constant and ripple voltage
  - $\tau\uparrow$ (*R*$\uparrow$ or *C*$\uparrow$) $\rightarrow$ ripple voltage $\downarrow$
  - $\tau\uparrow$ (*R*$\downarrow$ or *C*$\downarrow$) $\rightarrow$ ripple voltage $\uparrow$

Output frequency
- A filtered full-wave rectifier voltage has much less ripple than a half-wave rectifier voltage.
- The output frequency of a full-wave rectifier is twice that of a half-wave rectifier.
  - Period (*T*): $\qquad T_{\text{Half}} = 2\, T_{\text{Full}}$

  - Frequency (*f*): $\quad f = \dfrac{1}{T}, f_{\text{Full}} = \dfrac{1}{T_{\text{Full}}} = \dfrac{1}{\frac{1}{2}T_{\text{Half}}} = 2 f_{\text{Half}}$

Ripple factor

- Ripple factor (RF):

  $$\text{Ripple factor} = \frac{\text{AC component of output (RMS value)}}{\text{DC component of output}} \quad \text{or} \quad RF = \frac{V_{AC}}{V_{DC}} = \frac{I_{AC}}{I_{DC}}$$

- $RF\downarrow \rightarrow$ better filter
- $\tau\uparrow$ ($R\uparrow$ or $C\uparrow$) $\rightarrow RF \downarrow \rightarrow$ ripple voltage $\downarrow$

Surge current

- Surge current (inrush current or switch-on surge): the instantaneous rise current in a short period of time in a power supply at turn-on.
- Filter capacitor and surge current: when a power supply is initially turned on, charging filter capacitor can result in a surge current.
- Use surge-limiting resistor: a resistor in series with the diode can be used to limit surge current.
- Minimum surge resistor value = $\dfrac{\text{Peak secondary voltage} - 2\,V_{BR}}{\text{Diode max forward surge current (from data sheet)}}$

  or $\quad R_{surge\dagger} = \dfrac{V_{pk(sec)} - 2\,V_{BR}}{I_{FSM}}$

Diode clipping/clamping circuits

- Clipper (limiter): a circuit that limits or clips off a portion of an input AC signal above or below certain levels.
- Clamper: a circuit that adds a DC level to an AC signal and moves the whole signal up or down.
- Clamper vs. clipper
  - Clipper: used to remove (limit) a portion of an AC signal. (The voltage that is clipped by clipper can change in shape _ an AC signal + a DC level.)
  - Clamper: used to add (shift) a DC level to an input AC signal. (It moves the whole signal up or down.)

Table 3.1 Clippers and clampers

Name	Type	Circuit	Formula
Clipper (limiter)	Positive clipper	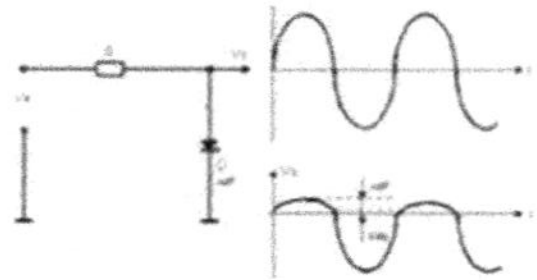	– Positive half-cycle: $V_{out} = V_{BR}$ – Negative half-cycle: $V_{out} = V_{in}$ – V across the load: $$V_{out} = V_{in}\frac{R_L}{R_L + R}$$
	Negative clipper	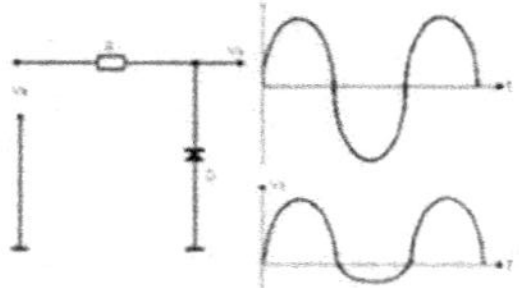	– Positive half-cycle: $V_{out} = V_{in}$ – Negative half-cycle: $V_{out} = -V_{BR}$ – V across the load: $$V_{out} = V_{in}\frac{R_L}{R_L + R}$$
	Biased positive clipper	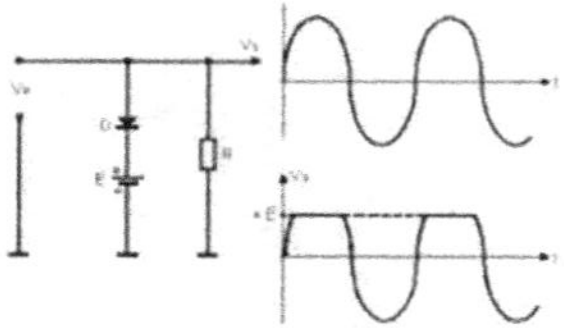	– Positive half-cycle: $V_{out} = E + V_{BR}$ – Negative half-cycle: $V_{out} = V_{in}$
	Biased negative clipper	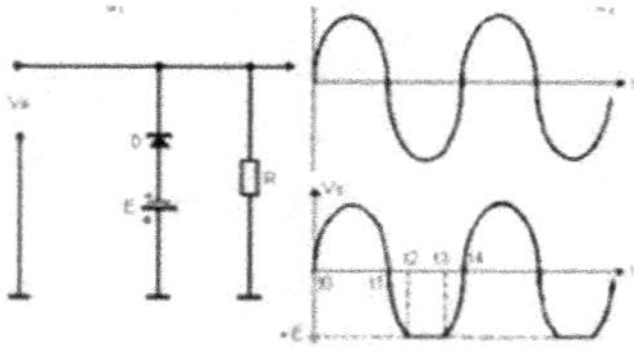	– Positive half-cycle: $V_{out} = V_{in}$ – Negative half-cycle: $V_{out} = -E - V_{BR}$

Combination clipper	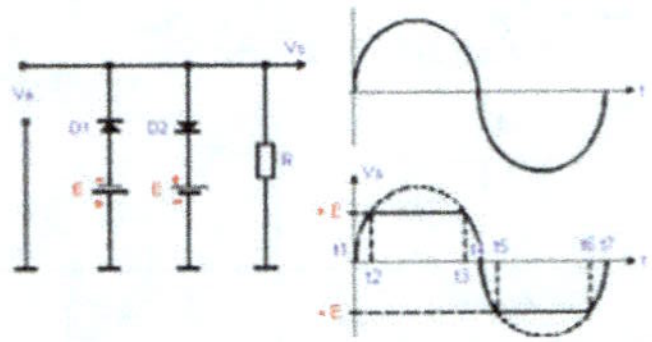	– Positive half-cycle: $$V_{out} = E + V_{BR}$$ – Negative half-cycle: $$V_{out} = -E - V_{BR}$$
Positive clamper	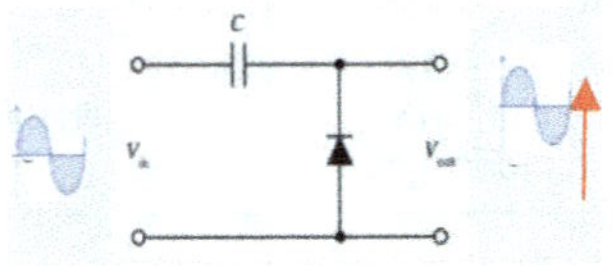	– Negative half-cycle: $$V_C = V_{pk(in)} - V_{BR}$$ – Positive half-cycle: $$V_{pk(out)} = V_C + V_{pk(in)}$$
Negative clamper	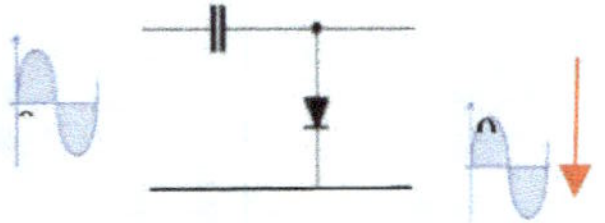	– Positive half-cycle: $$V_C = V_{pk(in)} - V_{BR}$$ – Negative half-cycle: $$V_{pk(out)} = -V_C + V_{pk(in)}$$
Biased clamper	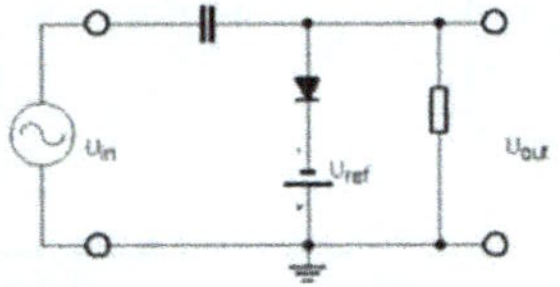	

Clamper

Table 3.2 *Rectifiers*

	Half-wave	Transformer-coupled half-wave	Full-wave center-tapped	Full-wave bridge
Circuit	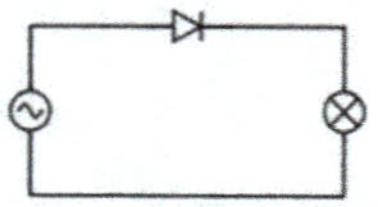	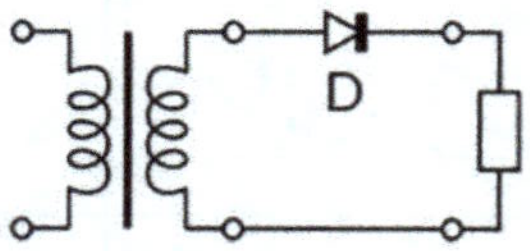	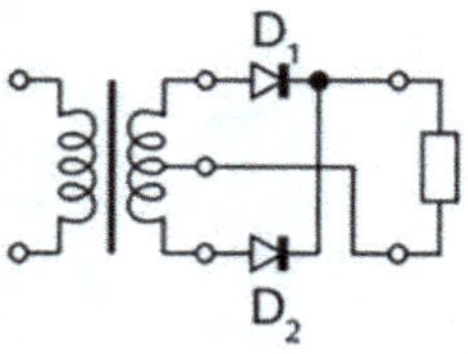	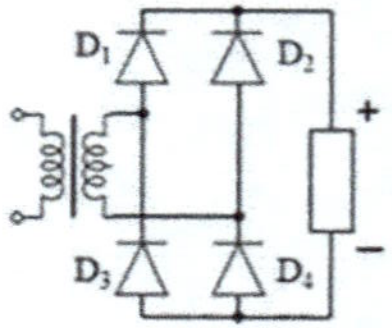
Waveform	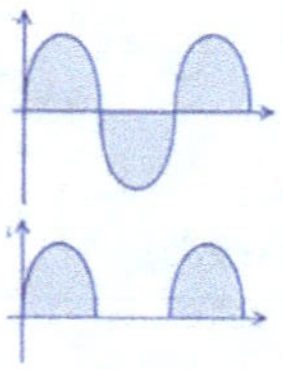	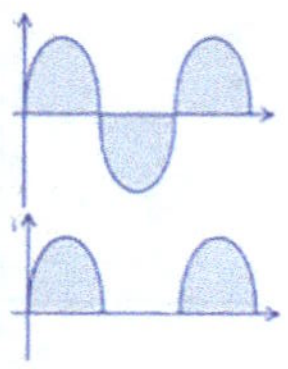	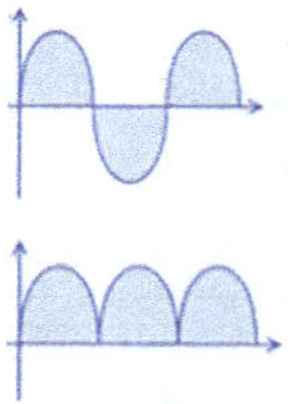	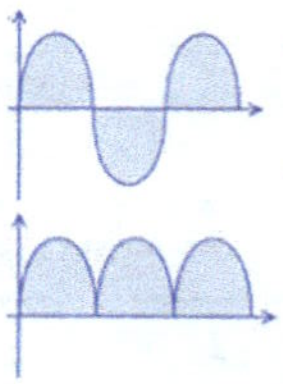
Operation	(+) half-cycle: D – FB (–) half-cycle: D – RB	(+) half-cycle: D – FB (–) half-cycle: D – RB	(+) half-cycle: D_1 – FB D_2 – RB (–) half-cycle: D_2 – FB D_1 – RB	(+) half-cycle: $D_1\,D_4$ – FB $D_2\,D_3$ – RB (–) half-cycle: $D_2\,D_3$ – FB $D_1\,D_4$ – RB
$V_{pk(out)}$	$V_{pk(out)} = V_{pk(in)} - V_{BR}$	$V_{pk(out)} = V_{pk(sec)} - V_{BR}$	$V_{pk(out)} = \dfrac{1}{2}V_{sec} - V_{BR}$	$V_{pk(out)} = V_{pk(sec)} - 2\,V_{BR}$
V_{avg}	$V_{avg} = \dfrac{V_{pk(out)}}{\pi}$ $V_{avg} \approx 0.318\,V_{pk(out)}$	$V_{avg} = \dfrac{V_{pk(out)}}{\pi}$ $V_{avg} \approx 0.318\,V_{pk(out)}$	$V_{avg} = \dfrac{2V_{pk(out)}}{\pi}$ $V_{avg} \approx 0.638\,V_{pk(out)}$	$V_{avg} = \dfrac{2V_{pk(out)}}{\pi}$ $V_{avg} \approx 0.638V_{pk(out)},$
PIV	$PIV = V_{pk\,(in)}$	$PIV = V_{pk(sec)}$	$PIV = V_{pk(sec)} \square\,V_{BR}$	$PIV = V_{pk\,(out)} + V_{BR}$

Self-test

3.1 1. A basic DC power supply consists of four basic circuits, i.e. transformer, (   ), filter, and regulator.

2. A voltage (   ) is a circuit that is designed to maintain the output voltage to a constant level despite changes in its input voltage or load.

3. A (   ) is most commonly used to raise or reduce the output voltage of the circuits.

4. A (   ) transformer is designed to provide two separate and equal secondary voltages.

5. The peak inverse voltage (PIV) is the maximum voltage a diode can withstand in the (   )-biased direction.

6. Determine the peak value of the output voltage and the PIV for Figure 3.41, if the turns ratio of the transformer is 0.5 and the peak input voltage is 110 V.

(A silicon diode)

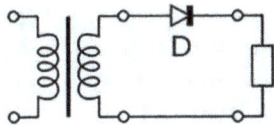

Figure 3.42   Ch 3 - No. 6, self-test

3.2 7. The output voltage of a center-tapped full-wave rectifier is always one-half of the total (   ) voltage less the diode drop.

8. Determine the peak secondary voltage, the peak output voltage, the PIV, and the average output voltage for the circuit in Figure 3.42, if the turns ratio of the transformer is 2 (1:2) and the peak input voltage is 120 V. (Assuming two silicon diodes)

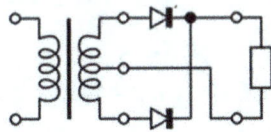

Figure 3.43   Ch 3 - No. 8, self-test

9. Determine the peak output voltage and the PIV for the bridge rectifier circuit in Figure 3.44, if the turns ratio of the transformer is 2 (1:2) and the peak input voltage is 120 V. (Four silicon diodes)

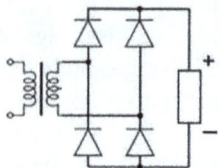

Figure 3.44   No. 9, self-test

3.3 10. In power supplies, filters are used to smooth out the pulsating (   ) output obtained from the rectifier circuit.
11. A capacitor (   ) can be used to reduce the AC components and make the output of rectifier smooth.
12. The (   ) of the *R* and *C* is called the *RC* time constant.
13. High *RC* time constant τ leads to a (   ) charging or discharging time.
14. High *RC* time constant τ leads to a (   ) ripple of the output voltage.
15. The (   ) factor is the ratio of the AC component's RMS to the DC component of the output voltage.
3.4 16. The (   ) is a circuit that limits or clips off a portion of an input AC signal above or below certain levels.
17. A negative clipper shown in Figure 3.45 has an RMS input voltage of 6.5 V, R_1 is 4 kΩ, and the load resistor is 40 kΩ. What is the peak output voltage for the clipper? (Assuming a silicon diode)

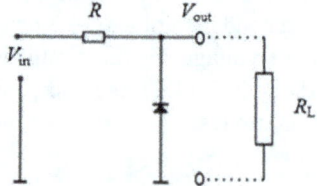

Figure 3.45   Ch 3 - No. 17, self-test

18. A (   ) clipper is a clipper circuit that uses a DC voltage source to adjust the limiting level of the output voltage.
19. A biased (   ) clipper is a clipper circuit that adjusts the negative half-cycles of the limiting level.

20. A (   ) clipper is a clipper circuit that adjusts both the positive and negative half-cycles of the limiting level.
21. A (   ) is a circuit that adds a DC level to an AC signal and moves the whole signal up or down.
22. A negative clamper shown in Figure 3.46 has a peak input voltage of 10 V. What is the RMS output voltage for the clamper? (Assuming a silicon diode)

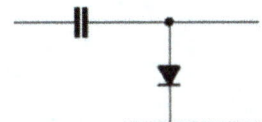

Figure 3.46   Ch 3 - No. 22, self-test

23. A biased clamper is a clamper circuit that uses a (   ) voltage source to have an additional shift level of the output voltage.

Chapter 4

Bipolar junction transistors

Chapter outline

4.1 Introduction to bipolar junction transistor

4.1.1 Bipolar junction transistor (BJT)

Introduction to transistors

- Transistor: a three-terminal semiconductor component that can work either to amplify electrical signals (amplifier) or in a high-speed switching circuit (switch).
- Types of transistors
  - Bipolar junction transistor (BJT): a current-controlled transistor that uses both electrons and holes charge carriers to conduct current.
  - Field-effect transistor (also known as unipolar transistor): a voltage-controlled transistor that uses electrons or holes charge carriers in their operation.

Types of BJTs

- BJTs are available in PNP and NPN types.
- NPN transistor: a transistor in which one P-type semiconductor layer is placed between two N-type semiconductor layers, and electrons are the majority charge carriers.

Unlike a normal P–N junction diode, the transistor has two P–N junctions.

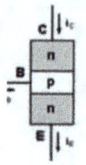   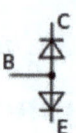   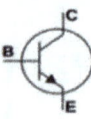

Figure 4.1(a)   NPN bipolar   Figure 4.1(b)   Two   Figure 4.1(c)   NPN
junction transistor   P–N junctions   BJT symbol

- PNP transistor: a transistor in which one N-type semiconductor layer is placed between two P-type semiconductor layers, and holes are the majority charge carriers.

The direction of the "arrow" specifies the type of BJT.

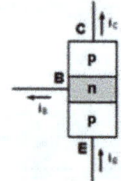   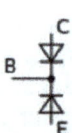   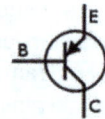

Figure 4.2(a)   PNP   Figure 4.2(b)   Two   Figure 4.2(c)   PNP
bipolar junction transistor   PN junctions   BJT symbol

BJT – basic structure

- A BJT has three layers (three regions)
  - Collector
  - Base
  - Emitter

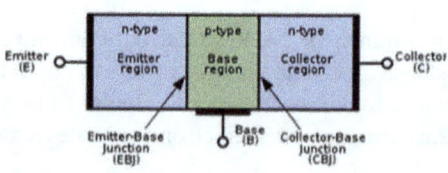

Figure 4.3   NPN BJT structure

- A BJT has two P–N junctions:
  - Base – emitter junction
  - Base – collector junction

- The base region is very thin and lightly doped.

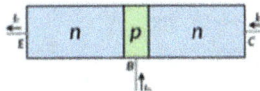

Figure 4.4   The base region is very thin

BJT biasing
- Biasing: establishing predetermined voltages or currents in an electronic circuit.
- BJT bias: to operate a BJT as an amplifier
    - the B–E junction should be forward biased ($V_{BE} \approx V_{BR}$).

    $V_{BR} = 0.7$ V for Si or 0.3 V for Ge

    - the C–B junction should be reverse biased.

BJT currents
- Emitter current I_E: the forward bias causes the electrons in the emitter to flow toward the base.
- Baser current I_B: the base is thin and lightly doped, only a few of the electrons from the emitter into the base will combine with limited holes. This produces the very small base current I_B.
- Collector current I_C: most of the electrons entering the base region will flow toward the collector.

    The B–C junction is designed to handle a large amount of reverse current without damage to the junction (it behaves as a Zener diode).

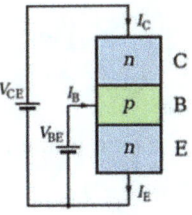

Figure 4.5   BJT currents

Emitter current
- NPN transistor: $I_E = I_C + I_B$ KCL: $\sum I_{in} = \sum I_{out}$

    $I_E \approx I_C$ I_B is very small.
- PNP transistor: $I_E = I_C + I_B$

    $I_E \approx I_C$

Figure 4.6a   NPN transistor   *Figure 4.6b   PNP transistor*

4.1.2   Transistor parameters

Alpha and beta parameters

- Signal gain (amplification factor): the amplified difference between the input and output AC signals.
  - It is a measure of the amount an amplifier "amplifies" the input signal.
  - It is the ratio of the output signal and the input signal: $\text{Gain} = \dfrac{\text{output}}{\text{input}}$

- Transistor alpha (α) and beta (β) parameters: the current gains of a BJT transistor.
  - Alpha (α): the ratio of the collector current to the emitter current. It is always less than 1.

$$\alpha = \frac{\text{collector current}}{\text{emitter current}} < 1$$

The collector current is the output, and the emitter current is the input.

  - Beta (β): the ratio of the emitter current to the base current. It is always great than 1.

$$\beta = \frac{\text{emitter current}}{\text{base current}} > 1$$

The emitter current is the output, and the base current is the input.

DC current gain of a transistor

- DC current gain alpha (α_{DC}): the ratio of the DC collector current I_C to the DC emitter current I_E.

$$\alpha_{DC} = \frac{I_C}{I_E}$$

- DC current gain beta (β_{DC}): the ratio of the DC collector current I_C to the DC base current I_B (it is also called collector efficiency). $\beta_{DC} = \dfrac{I_C}{I_B}$

- Hybrid parameter (h_{FE}): it is the same as the DC beta β_{DC} but is more widely used in transistor datasheets.

h_{FE}:   F – forward bias;   E – common emitter configuration

Example: A transistor operates with a base current of 2 mA and an emitter current of 200 mA. Determine the DC current gains beta (β_{DC}) and alpha (α_{DC}).

- Given: $I_B = 2$ mA and $I_E = 200$ mA.
- Find: α_{DC} and β_{DC}.
- Solution:

$$\alpha_{DC}: \quad \alpha_{DC} = \frac{I_C}{I_E} = \frac{I_E - I_B}{I_E} = \frac{200 \text{ mA} - 2 \text{ mA}}{200 \text{ mA}} = 0.99 \quad (<1) \quad I_E = I_C + I_B, \; I_C = I_E - I_B$$

$$\beta_{DC}: \quad \beta_{DC} = \frac{I_C}{I_B} \approx \frac{I_E - I_B}{I_B} = \frac{200 \text{ mA} - 2 \text{ mA}}{2 \text{ mA}} = 99 \quad (>1)$$

The relationship between alpha (α_{DC}) and beta (β_{DC})

- α_{DC}: $\alpha_{DC} = \dfrac{\beta_{DC}}{\beta_{DC} + 1}$

     Derive: $I_E = I_C + I_B,$ $\dfrac{I_E}{I_C} = \dfrac{I_C}{I_C} + \dfrac{I_B}{I_C}$ Divide by I_C

             $\dfrac{I_E}{I_C} = 1 + \dfrac{I_B}{I_C},$ $\dfrac{1}{\alpha_{DC}} = 1 + \dfrac{1}{\beta_{DC}}$ $\alpha_{DC} = \dfrac{I_C}{I_E}, \quad \beta_{DC} = \dfrac{I_C}{I_B}$

             $\dfrac{1}{\alpha_{DC}} = \dfrac{\beta_{DC} + 1}{\beta_{DC}},$ $\alpha_{DC} = \dfrac{\beta_{DC}}{\beta_{DC} + 1}$

- β_{DC}: $\beta_{DC} = \dfrac{\alpha_{DC}}{1 - \alpha_{DC}}$

     Derive: $\alpha_{DC} = \dfrac{\beta_{DC}}{\beta_{DC} + 1}$

           $\alpha_{DC}(\beta_{DC} + 1) = \beta_{DC}$

           $\alpha_{DC}\beta_{DC} + \alpha_{DC} = \beta_{DC}$

           $\alpha_{DC} = \beta_{DC} - \alpha_{DC}\beta_{DC}$ Subtract $\alpha_{DC}\beta_{DC}$

           $\alpha_{DC} = \beta_{DC}(1 - \alpha_{DC})$ Factor out β_{DC}

           $\beta_{DC} = \dfrac{\alpha_{DC}}{1 - \alpha_{DC}}$ Divide by $(1 - \alpha_{DC})$

Example: A transistor operates with a base current of 70 μA and a collector current of 6 mA. Determine the DC current gains beta β_{DC} and alpha α_{DC}.

- Given: $I_B = 70\ \mu A$, $I_C = 6\ mA$
- Find: β_{DC} and α_{DC}
- Solution:

$\boldsymbol{\beta_{DC}}$: $\beta_{DC} = \dfrac{I_C}{I_B} = \dfrac{6\ mA}{70\ \mu A} = \dfrac{6\ mA}{0.07\ mA} \approx \mathbf{85.7}$ Milli: 10^{-3}; Micro (μ): 10^{-6}

$\boldsymbol{\alpha_{DC}}$: $\alpha_{DC} = \dfrac{\beta_{DC}}{\beta_{DC} + 1} = \dfrac{85.7}{85.7 + 1} \approx \mathbf{0.989}$

Beta curve

- β_{DC} and temperature: the DC current gain (β_{DC}) is influenced not only by collector current (I_C) but also by temperature (T).
- Beta varies with both:
  - junction temperature (T)
  - collector current (I_C)
- I_C plots have different curves for different temperatures.

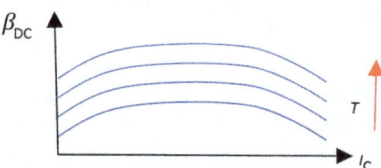

Figure 4.7   Beta curves

The relationship between I_C, T, and β_{DC}

- If the collector current (I_C) increases (the junction temperature T does not change), the DC current gain (β_{DC}) will increase. But when β_{DC} reaches the maximum value, the increasing collector current (I_C) will cause the DC current gain (β_{DC}) decrease.

$$I_C \uparrow \left(\vec{T} \right) \rightarrow \beta_{DC} \uparrow \rightarrow \text{maximum} \rightarrow I_C \uparrow \rightarrow \beta_{DC} \downarrow$$

- If the junction temperature (T) increases (the collector current I_C does not change), the DC current gain (β_{DC}) will increase.

$$T \uparrow \left(\vec{I_C} \right) \rightarrow \beta_{DC} \uparrow$$

Maximum ratings of transistors

- Maximum ratings: the upper limit of usage conditions or the highest values that must not be exceeded during operation.
- Maximum transistor ratings: the maximum allowable current, voltage, power dissipation, etc.

  Understanding maximum ratings are critical to the reliable operation of transistors.
- Maximum transistor ratings in the datasheet:
  - Maximum DC voltage across the collector base: V_{CBO} $V_{CBO} = V_{CB\,(max)}$
  - Maximum DC voltage across the collector emitter: V_{CEO} $V_{CEO} = V_{CE\,(max)}$
  - Maximum DC voltage across the base emitter: V_{BEO} $V_{BEO} = V_{BE\,(max)}$

    Voltages given in the V--o form are usually maximum ratings.
  - Maximum DC collector current: $I_{C\,(max)}$
  - Maximum DC power dissipation: $P_{D\,(max)}$

    The product of I_C and V_{CE} cannot be maximum at the same time. $P = I\,V$

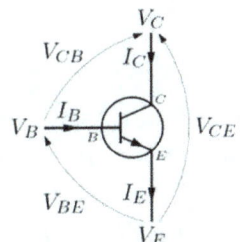

Figure 4.8   BJT voltages and currents

4.2   DC analysis of a BJT circuit

4.2.1   DC analysis

DC voltages and currents

- DC analysis: analyzes and calculates the behavior of a circuit when the circuit is connected only with DC sources.
- Common emitter configuration: the emitter (E) of the transistor is common in both input (base – emitter) and output circuit (collector – emitter).

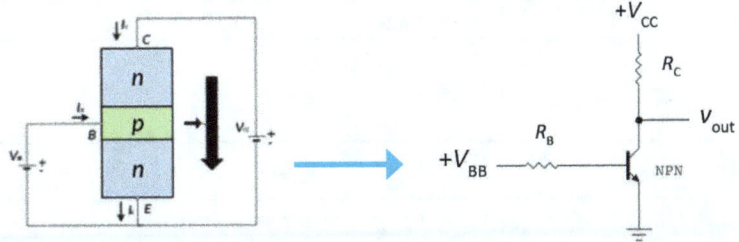

*Figure 4.9(a)   Common emitter
configuration*

*Figure 4.9(b)   Common emitter
circuit*

Table 4.1   DC voltages and currents in a BJT circuit

DC current		DC voltage	
Base current:	I_B	Base – emitter voltage:	V_{BE}
Emitter current:	I_E	Collector – base voltage:	V_{CB}
Collector current:	I_C	Collector – emitter voltage:	V_{CE}

Calculating DC voltages and currents

- Base current I_B:

$$I_B = \frac{V_{BB} - V_{BE}}{R_B}$$

    Derive: $V_{RB} = V_{BB} - V_{BE}$ V_{RB}: the voltage across R_B

$$I_B R_B = V_{BB} - V_{BE}$$

$$I_B = \frac{V_{BB} - V_{BE}}{R_B}$$ Divide both sides by R_B

- Emitter current I_E:

$$I_E = I_C + I_B \approx I_C \quad \text{or} \quad I_E = \frac{I_C}{\alpha_{DC}} \qquad \alpha_{DC} = \frac{I_C}{I_E}$$

- Collector current I_C: $I_C = \beta_{DC} I_B$ $\beta_{DC} = \frac{I_C}{I_B}$

- Base-emitter voltage V_{BE}: $V_{BE} \approx V_{RB}$ $V_{BR} = 0.7$ V for Si or 0.3 V for Ge

- Collector-emitter voltage V_{CE}: $V_{CE} = V_{CC} - I_C R_C$ KVL: $V_{CC} = I_C R_C + V_{CE}$

- Collector-base voltage V_{CB}: $V_{CB} = V_{CE} - V_{BE}$ KVL: $V_{CE} = V_{CB} + V_{BE}$

Table 4.2   DC currents and voltages

DC current	DC voltage
I_B: $I_B = \dfrac{V_{BB} - V_{BE}}{R_B}$	V_{CE}: $V_{CE} = V_{CC} - I_C R_C$
I_C: $I_C = \beta_{DC} I_B$	V_{CB}: $V_{CB} = V_{CE} - V_{BE}$
I_E: $I_E = I_C + I_B \approx I_C$	V_{BE}: $V_{BE} \approx V_{BR} = 0.7$ V (Si.)
	$V_{BE} \approx V_{BR} = 0.3$ V (Ge)

DC analysis of a BJT circuit – an example

Example: A common emitter circuit has an R_B of 12 kΩ, R_C of 150 Ω, supply voltage V_{BB} of 4 V, V_{CC} of 12 V, and β_{DC} of 130. Determine the DC currents I_B, I_C, I_E, and the DC voltages V_{CE} and V_{CB}. (Assuming a silicon transistor)

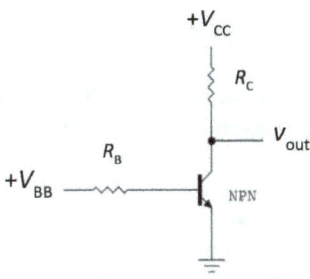

Figure 4.10   Common emitter – an example

- Given: $R_B = 12$ kΩ, $R_C = 150$ Ω, $V_{BB} = 4$ V, $V_{CC} = 12$ V, $\beta_{DC} = 130$
- Find: I_B, I_C, I_E, V_{CE}, and V_{CB}.
- Solution:

I_B: $$I_B = \frac{V_{BB} - V_{BE}}{R_B} = \frac{4\ V - 0.7\ V}{12\ k\Omega}$$ $V_{BE} \approx V_{BR} = 0.7$ V

$$= 0.275\ mA = 275\ \mu A$$ Milli: 10^{-3}; Micro(μ) : 10^{-6}

I_C: $I_C = \beta_{DC} I_B = (130)\ (275\ \mu A) = 35{,}750\ \mu A = 35.75\ mA$

I_E: $I_E = I_C + I_B = 35{,}750\ \mu A + 275\ \mu A = 36{,}025\mu A = 36.025\ mA$ $I_E \approx I_C$

V_{CE}: $V_{CE} = V_{CC} - I_C R_C = 12\ V - (35.75\ mA)\ (150\ \Omega)$

$$= 12\ V - 5365.2\ mV \approx 12\ V - 5.3652\ V \approx 6.635V$$

V_{CB}: $V_{CB} = V_{CE} - V_{BE} = 6.635\ V - 0.7\ V = 5.935\ V$

4.2.2   BJT characteristic curves

BJT curves

- Transistor characteristics: curves which are drawn between currents and voltages of a transistor in the given configuration.
- Input and output characteristic curves of a transistor:
  - Input characteristics (base curve): the graph of the base current I_B versus the emitter-base voltage V_{BE} (for specified values of the collector-emitter voltage V_{CE}).

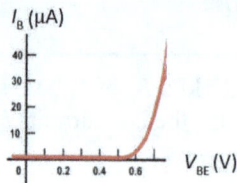

Figure 4.11(a)   Input characteristics

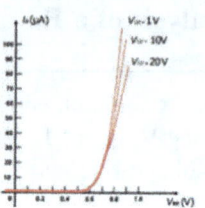

Figure 4.11(b)   Base curves

- Output characteristics (collector curve): the graph of the collector current I_C versus the collector-emitter voltage V_{CE} (for specified values of the base current I_B).

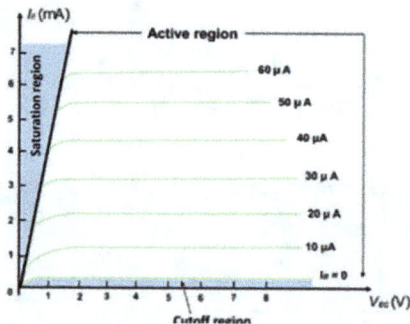

Figure 4.12   Output characteristics

A family curves of I_C versus V_{CE}

Family of collector characteristic curves
- Family of collector curves: I_C versus V_{CE} for different values of the base current I_B.
- Change in V_{BB} causes a change in I_B. $V_{BB} \rightarrow I_B$
- Each curve in this family shows the dependence of I_C on V_{CE} when I_B has a constant value.

Generate collector curves
- Collector curves can be generated by using a circuit shown in Figure 4.13 (both V_{BB} and V_{CC} are adjustable source voltages).
- When $V_{CC} = 0$: $V_{CE} = 0$, $I_C = 0$

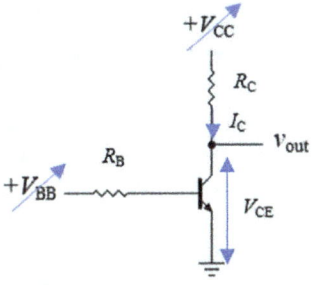

Figure 4.13 Generate collector curves

- When V_{CC} is increased, I_C increases gradually as V_{CE} increases.

  When $V_{CC} \uparrow$: $\qquad\qquad V_{CE} \uparrow \rightarrow I_c \uparrow$

- When V_{CE} exceeds V_{BR} (such as 0.7 V for Si), the B–C junction becomes reverse biased, I_C remains constant (for a given value of I_B) as V_{CE} continues to increase.

  When $V_{CE} \geq V_{BR}$: $\qquad\qquad V_{CE} \overset{\uparrow}{\rightarrow} \overline{I_C} \qquad\qquad \overline{I_C} = \overline{\beta_{DC} \; I_B}$

- When V_{CE} reaches a sufficiency high value, the reverse-biased B–C junction goes into breakdown, and I_C increases shapely.

  When $V_{CE} \uparrow \geq V_{CE \, (max)} \rightarrow$ breakdown $\rightarrow I_c \uparrow$

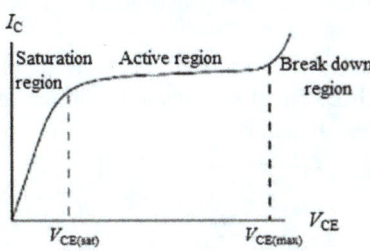

Figure 4.14 Collector curve

BJT – switch and amplifier

- The transistors have two basic functions: "switching" (switch) or "amplification" (amplifier).
- The transistors can operate in three different regions: the cutoff region, saturation region, and active region.

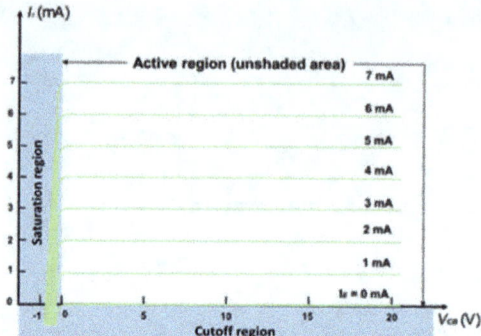

Figure 4.15   Three regions

4.2.3   Distinct regions of collector curves

BJT regions of collector curves

- Cutoff region: both B–E and B–C junctions are reverse biased and have almost no current flows (only very small reverse leakage current). The BJT acts as an open switch.

$$V_{BB} = 0, \qquad I_B \approx 0 \qquad (I_B = I_{BCO}) \qquad\qquad V_{BE} < V_{RB} \text{ (RB)}$$

I_{BCO} is a very small reverse leakage current between C and B (due to thermally produced minority carriers).

  - $I_C \approx 0 \qquad (I_C = I_{CEO})$
    I_{CEO} is a very small reverse leakage current between the collector and the emitter.
  - BJT ≈ an open switch

- Saturation region: both B–E junction and C–B junction are forward biased. High I_C flows through BJT and reaches a maximum level (it hits a limit). The transistor acts as a closed switch.

  - $V_{CE} < V_{CE(sat)}$ or V_{BR}: $I_B \uparrow \to I_C \uparrow$ $\qquad I_C = \beta_{DC} I_B, I_B$ is dependent on I_B and β_{DC}.
    $V_{CE(sat)}$: the saturation voltage between the collector and emitter terminals.
  - $V_{CE} = V_{CE(sat)}$ or V_{BR}: $I_B \uparrow \to \overline{I_C} = I_{C(sat)}$ $\qquad I_C = \beta_{DC} I_B$ no longer holds true.
    ○ $I_{C(sat)}$: the collector saturation current
    ○ When V_{CE} reaches its saturation value $V_{CE(sat)}$, I_C can increase no further even with a continued increase in I_B.
  - BJT ≈ a closed switch

- Active (linear) region: the B–E junction is forward biased, and the C–B junction is reverse biased. I_C is β times I_B. The transistor acts as an amplifier.

  - $V_{CE} > V_{BR}$: $V_{CC} \uparrow \to V_{CE} \uparrow \to I_C$ slightly $\uparrow$ $\qquad I_C$ slightly increases at certain I_B.
  - $I_B \uparrow$: $I_B \uparrow \to I_C \uparrow$ $\qquad\qquad I_C = \beta_{DC} I_B, I_B$ controls I_C – a family curves

$$V_{CE(sat)} < V_{CE} < V_{CE(max)}$$

$V_{CE(max)}$: the maximum voltage between the collector and emitter.

- Breakdown region: V_{CC} is so large that the C–B junction breaks down, causing a large I_C to flow and damage to the transistor.

$$V_{CC} \uparrow\uparrow \quad \rightarrow \quad V_{CE} \uparrow\uparrow \geq V_{CE\,(max)} \quad \rightarrow I_C \uparrow\uparrow \rightarrow \text{breakdown}$$

Table 4.3   BJT regions of operation

Region	Bias	Condition	I_C	Acts as
Cutoff	C–B: RB; B–E: RB	$V_{BE} < V_{RB}$, $I_B \approx 0$, $I_C \approx 0$ $(I_B = I_{BCO},\ I_C = I_{CEO})$	$I_C \approx 0$	An open switch (off)
Saturation	C–B: FB; B–E: FB	$V_{CE} < V_{BR}$: $I_B \uparrow \rightarrow I_C \uparrow$ $V_{CE} = V_{BR}$: $I_B \uparrow \rightarrow \uparrow \rightarrow \overrightarrow{I_C}$	$I_C = I_{C\,(sat)}$	A closed switch (on)
Active	C–B: RB; B–E: FB	$I_B \uparrow \rightarrow I_C \uparrow$	$I_C = \beta_{DC} I_B$	An amplifier

Transistor DC load line analysis

- DC load line: a line drawn on the collector characteristic curve by connecting the cutoff and saturation points.

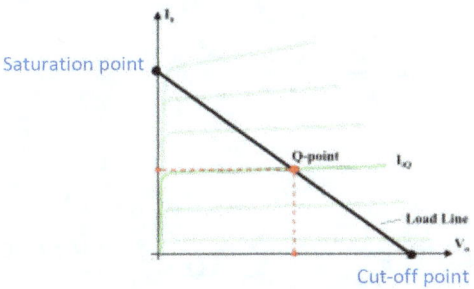

Figure 4.16   DC load line

- Saturation point: a point at which corresponds to the maximum possible value of I_C and the minimum V_{CE} value.

$$V_{CE} = V_{CE\,(sat)}, \qquad I_C = I_{C(max)}$$

- Cutoff point: a point at which corresponds to the maximum possible value of V_{CE} and the minimum I_C value.

$$V_{CE} = V_{CE\,(max)} \approx V_{CC}, \qquad I_C \approx 0 \qquad\qquad V_{CE} = V_{CC} - I_C R_C, I_C \approx 0$$

- The region between cutoff and saturation is known as an active region.
- The DC load line is the locus of all possible operating points at which BJT remains in the active region.

Example: Determine if the BJT in Figure 4.17 is saturated for the following values: R_2 is 20 kΩ, R_1 is 3 kΩ, V_{BB} is 4 V, V_{CC} is 12 V, β_{DC} is 50, and $V_{CE(max)}$ is 0.1 V. (Assuming a silicon transistor)

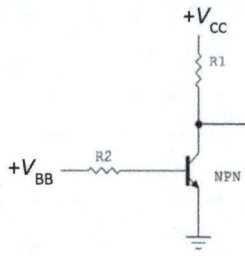

Figure 4.17   Common emitter – saturated?

- Given: $R_B = 20$ kΩ, $R_C = 3$ kΩ, $V_{BB} = 4$ V, $V_{CC} = 12$ V, $\beta_{DC} = 50$, $V_{CE(sat)} = 0.1$ V

  $$R_B = R_2, R_C = R_1$$

- Find: $I_{C(sat)}$ and I_C (Check I_C, if $I_C > I_{C(sat)} \rightarrow$ it is saturated)

- Solution: $I_C(sat) = \dfrac{V_{CC} - V_{CE(sat)}}{R_C}$; $I_C \overset{?}{=} \beta_{DC} I_B \ \leftarrow \ I_B = \dfrac{V_{BB} - V_{BE}}{R_B}$

I_C (sat): $I_C(sat) = \dfrac{V_{CC} - V_{CE(sat)}}{R_C}$ $V_{CE} = V_{CC} - I_C R_C$

$$= \dfrac{12 \text{ V} - 0.1 \text{ V}}{3 \text{ k}\Omega} \approx \textbf{3.97 mA}   \text{Kilo: } 10^3; \text{ Milli: } 10^{-3}$$

I_B: $I_B = \dfrac{V_{BB} - V_{BE}}{R_B} = \dfrac{4 \text{ V} - 0.7 \text{ V}}{20 \text{ k}\Omega} = 0.165 \text{ mA}$ $V_{BB} = I_B R_B + V_{BE}$

I_C: $I_C = \beta_{DC} I_B = (50) (0.165 \text{ mA}) = \textbf{8.25 mA}$

Saturation? $I_C > I_{C(sat)} \ \rightarrow \ $ **Saturated**

4.3   Transistors as amplifiers and switches

4.3.1   Transistor as an amplifier

BJT amplifier

- The BJT acts as amplifiers and switches
  - A transistor in the active (linear) region can work as an amplifier.
  - A transistor in the saturation or cutoff regions can work as a switch.

- Amplifier: an electronic device that produces an output that is larger than the input. It increases the amplitude of signals (voltage, current, power, etc.).

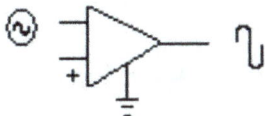

Figure 4.18   Amplifier

- DC and AC quantities: a BJT amplifier circuit has both DC and AC quantities associated with it and has a combination of both DC and AC operation.

Voltage and current notation (in this book)

- DC quantities use capital letters with capital subscripts.

    Example: I_B, I_C, I_E, V_{CE}, V_{CB}, V_{BE}, ...

- AC quantities use lowercase letters with capital subscripts (except r_e).

    Example: i_B, i_C, i_E, v_{CE}, v_{BE}, v_{CB}, r_e, ...

- Instantaneous quantities are represented by lowercase letters with lowercase subscripts.

    Example: i_c, i_e, v_c, v_{ce}, ...

Table 4.4   Voltage and current notation

Quantity	Letter	Subscript	Example
DC	Capital	Capital	I_B, I_C, V_{CE}
AC	Lowercase	Capital (except r_e)	i_B, i_C, $v_{CE,}$ r_e
Instantaneous	Lowercase	Lowercase	i_c, i_e, v_c, v_{ce}

BJT AC analysis

- Transistor modeling: use an equivalent circuit (model) to represent the AC characteristics of the transistor.
- AC equivalent circuit: an AC equivalent circuit can be made by reducing all DC sources to zero (short circuit all DC voltage sources).

The r_e transistor model

- r_e model: an AC equivalent circuit that can be used to predict the performance of a BJT.

- r_e: the internal AC resistance looking into the emitter terminal of a BJT.

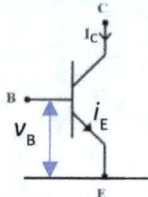

Figure 4.19 r_e

- Calculating r_e:

$$r_e = \frac{v_B}{i_E}$$

$$v_B = v_{in} - i_B R_B, \quad i_E \approx i_C, \quad i_C = \frac{v_C}{R_C}$$

Common emitter voltage gain

- Voltage gain (A_V): the ratio of the output voltage to the input voltage.

$$A_V = \frac{\text{Output voltage}}{\text{Input voltage}} = \frac{v_{out}}{v_{in}}$$

- The voltage gain in the common emitter configuration:

$$A_V = \frac{v_{out}}{v_{in}} = \frac{v_C}{v_B} = \frac{i_C R_C}{i_E r_e} \approx \frac{R_C}{r_e}$$

$$i_E \approx i_C$$

$$A_V \approx \frac{R_C}{r_e}$$

- The voltage gain A_V is dependent on the R_C and r_e.

Example: A common emitter circuit has r_e of 20 Ω, a 30 mV input signal at the base of the transistor, and an output voltage of 1.2 V at the collector. Determine the collector resistance R_C.

- Given: $r_e = 20\ \Omega, \quad v_{in} = 30\ \text{mV}, \quad v_{out} = 1.2\ \text{V}.$
- Find: R_C
- Solution: $R_C = A_V \overset{?}{r_e} \leftarrow A_V = \frac{v_{out}}{v_{in}}$

$$A_V \approx \frac{R_C}{r_e}$$

A_V: $A_V = \frac{v_{out}}{v_{in}} = \frac{1.2\ \text{V}}{30\ \text{mV}} = \frac{1,200\ \text{mV}}{30\ \text{mV}} = 40$

R_C: $R_C = A_V r_e$
$= (40)\ (20\ \Omega) = 800\ \Omega$

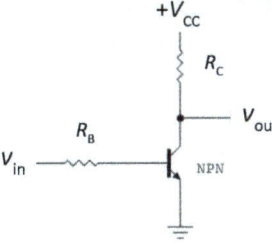

Figure 4.20   Common emitter: RC = ?

4.3.2   *Transistor switching circuit*

Cutoff region (switching off)

• A transistor as a switch: a transistor in saturation or cutoff regions can work as a switch to switch a low-voltage DC device (e.g., light-emitting diode (LED)) on or off.

• Cutoff: both B–E and B–C junctions are reverse biased and almost no current flows.

  – $V_{in} = 0$, $I_B \approx 0$, $I_C \approx 0$ $I_B = I_{BCO},\; I_C = I_{CEO}$

  – BJT ≈ an open switch (switching off)

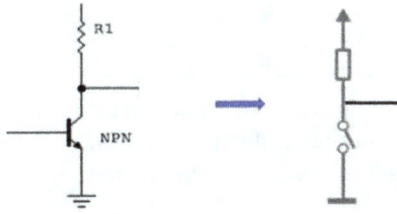

Figure 4.21   BJT switch

• Calculating cutoff voltage:

$$V_{CE\,(cutoff)} = V_{CE\,(max)} \approx V_{CC}$$

                                                             $V_{CE} = V_{CC} - I_C R_C,\;\; I_C \approx 0$

Saturation region (switching on)

• Saturation: both B–E junction and C–B junction are forward biased. I_C reaches a maximum.

  – $V_{in} \neq 0$

  – $V_{CE} < V_{CE(sat)}$ or V_{BR}: $I_B \uparrow \rightarrow I_C \uparrow$ $I_C = \beta_{DC} I_B$

  – $V_{CE} \geq V_{CE(sat)}$ or V_{BR}: $I_B \uparrow \rightarrow \overrightarrow{I_C} = I_{C\,(sat)}$

  – BJT ≈ α closed switch (switching on)

- Calculating saturation voltage and current:

$$V_{CE(sat)} \approx V_{BR}$$

$V_{BR} = 0.7\ V\ (Si)\ or\ 0.3\ V\ (Ge)$

$$I_{C(sat)} = \frac{V_{CC} - V_{CE\ (sat)}}{R_C}$$

$V_{CE} = V_{CC} = I_C R_C$

- The minimum base current $I_{B(min)}$ required to turn on a transistor:

$$I_{B(min)} = \frac{I_{C\ (sat)}}{\beta_{DC}}$$

$I_C = \beta_{DC} I_B$

Table 4.5   Cutoff and saturation

BJT	Switch	Voltage and current	Minimum/maximum current
Cutoff	Switch off (nonconducting)	$V_{in} = 0,\quad I_B \approx 0,\quad I_C \approx 0$ $I_B = I_{BCO},\quad I_C = I_{CEO}$	Minimum current $I_C \approx 0$
Saturation	Switch on (conducting)	$V_{in} \neq 0\quad (V_{in} \leq V_{BR})$ $V_{CE} < V_{BR}:\ I_B \uparrow \rightarrow I_C \uparrow$ $V_{CE} = V_{BR}:\ I_B \uparrow \rightarrow \overrightarrow{I_C}$	Maximum current $I_C = I_{C(sat)}$

Example: What the minimum value of input voltage is required to saturate the BJT in Figure 4.22, if the BJT has the following values: R_C is 1 kΩ, R_B is 50 kΩ, V_{BB} is 5 V, V_{CC} is 10 V, and β_{DC} is 75? (Assuming a silicon transistor)

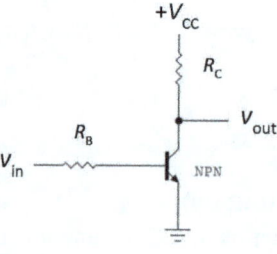

Figure 4.22   Common emitter – the min. input voltage?

- Given: $R_C = 1\ k\Omega$, $R_B = 50\ k\Omega$, $V_{BB} = 5\ V$, $V_{CC} = 10\ V$, and $\beta_{DC} = 75$.
- Find: $V_{in(min)}$ – the minimum V_{in} to saturate the BJT.

- Solution: Thinking process:

$$V_{in} = V_{BB} \overset{?}{=} I_B R_B + V_{BE}$$

$$\rightarrow I_{B(min)} = \frac{\overset{?}{I_{C\,(sat)}}}{\beta_{DC}} \rightarrow I_{C(sat)} = \frac{V_{CC} - V_{CE\,(sat)}}{R_C}$$

$I_{C(sat)}$:
$$I_{C(sat)} = \frac{V_{CC} - V_{CE\,(sat)}}{R_C} \approx \frac{V_{CC} - V_{BR}}{R_C} \qquad \qquad V_{CE(sat)} \approx V_{BR}$$

$$= \frac{10\ V - 0.7\ V}{1\ k\Omega} = 9.3\ mA \qquad \qquad \text{Kilo: } 10^3;\ \text{Milli: } 10^{-3}$$

$I_{B(min)}$:
$$I_{B(min)} = \frac{I_{C\,(sat)}}{\beta_{DC}} = \frac{9.3\ mA}{75} = 0.124\ mA$$

$V_{in(min)}$:
$$V_{in(min)} = V_{BB} = I_{B\,(min)}\,R_B + V_{BE} \qquad \qquad V_{BE} \approx V_{BR}$$

$$= (0.124\ mA)\,(50\ k\Omega) + 0.7\ V = \textbf{6.9 V}$$

The minimum value of input voltage is 6.9 V.

Basic application of BJT as a switch –
BJT driving an LED

- Turn on/off an LED with a switch
  - When the switch is off:
    $V_{in} = 0$: the transistor is in the cutoff and the LED is off. $V_{BB} \approx V_{in}$
  - When the switch is on:

    $V_{in} \neq 0$: the transistor is in saturation and the LED is on.

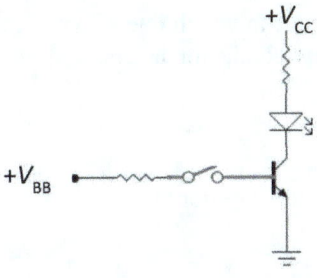

Figure 4.23   BJT as a switch

- Turn on/off an LED with a solar cell
  - When the power supply voltage (sunlight) is zero:
    $V_{in} = 0$: the transistor is in the cutoff and the LED is off.
  - When the power supply voltage (sunlight) increases:
    $V_{in} \uparrow$: the transistor is in saturation and the LED is on.

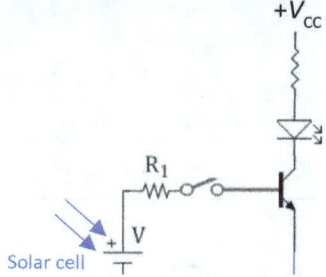

Figure 4.24   Turn on an LED with a solar cell

Summary

Bipolar junction transistor (BJT)

- Transistor: a three-terminal semiconductor component that can work either to amplify electrical signals (amplifier) or in a high-speed switching circuit (switch).
- Types of transistors
  - Bipolar junction transistor (BJT): a current-controlled transistor that uses both electrons and holes charge carriers to conduct current.
  - Field-effect transistor (FET – also known as unipolar transistor): a voltage-controlled transistor that uses electrons or holes charge carriers in their operation.

Types of BJTs

- NPN transistor: a transistor in which one P-type semiconductor layer is placed between two N-type semiconductor layers, and electrons are the majority charge carriers.
- PNP transistor: a transistor in which one N-type semiconductor layer is placed between two P-type semiconductor layers, and holes are the majority charge carriers.

BJT – basic structure

- A BJT has three layers (three regions)
  - Collector
  - Base
  - Emitter
- A BJT has two P–N junctions:
  - Base-emitter junction
  - Base-collector junction
- The base region is very thin and lightly doped

BJT bias

- Biasing: establishing predetermined voltages or currents in an electronic circuit.
- BJT bias: to operate a BJT as an amplifier
  - the B–E junction is forward biased ($V_{BE} \approx V_{BR}$). $V_{BR} = 0.7$ V for Si or 0.3 V for Ge
  - the C–B junction is reverse biased.

Alpha and beta parameters

- Signal gain: $\text{Gain} = \dfrac{\text{Output}}{\text{Input}}$

- Transistor alpha (α) and beta (β) parameters: the current gains of a BJT transistor.
  - Alpha (α): $\alpha = \dfrac{\text{collector current}}{\text{emitter current}} < 1$

  - Beta (β): $\beta = \dfrac{\text{emitter current}}{\text{base current}} > 1$

DC current gain of a transistor

- DC current gain alpha (α_{DC}): $\alpha_{DC} = \dfrac{I_C}{I_E}$

- DC current gain beta (β_{DC}): $\beta_{DC} = \dfrac{I_C}{I_B}$

- Hybrid parameter (h_{FE}): it is the same as the DC beta β_{DC} but is more widely used in transistor datasheets.

The relationship between α_{DC} and β_{DC}

- α_{DC}: $\alpha_{DC} = \dfrac{\beta_{DC}}{\beta_{DC} + 1}$

- α_{DC}: $\beta_{DC} = \dfrac{\alpha_{DC}}{1 - \alpha_{DC}}$

Beta curve

The relationship between I_C, T, and β_{DC}

- $I_C \uparrow (\vec{T}) \rightarrow \beta_{DC} \uparrow \rightarrow \text{maximum} \rightarrow I_C \uparrow\uparrow \rightarrow \beta_{DC} \downarrow$
- $T \uparrow (\vec{I_C}) \rightarrow \beta_{DC} \uparrow$

Maximum transistor ratings in the datasheet:

- Maximum DC voltage across the collector and base: V_{CBO} $V_{CBO} = V_{CB \, (max)}$
- Maximum DC voltage across the collector and emitter: V_{CEO} $V_{CEO} = V_{CE \, (max)}$
- Maximum DC voltage across the emitter and base: V_{EBO} $V_{EBO} = V_{EB \, (max)}$

- Maximum DC collector current: $I_{C\,(max)}$
- Maximum DC power dissipation: $P_{D\,(max)}$

DC voltages and currents

- DC analysis: analyzes and calculates the behavior of a circuit when the circuit is connected only with DC sources.
- Common emitter configuration: the emitter (E) of the transistor is common in both input (base-emitter) and output circuit (collector-emitter).

Table 4.6   DC voltages and currents in a BJT circuit

DC current		DC voltage	
Base current:	$I_B = \dfrac{V_{BB} - V_{BE}}{R_B}$	Base-emitter voltage:	$V_{BE} \approx V_{BR} = 0.7$ V (Si) $V_{BE} \approx V_{BE} = 0.3$ V (Ge)
Emitter current:	$I_E = I_C + I_B \approx I_C$	Collector-base voltage:	$V_{CB} = V_{CE} - V_{BE}$
Collector current:	$I_C = \beta_{DC} I_B$	Collector-emitter voltage:	$V_{CE} = V_{CC} - I_C R_C$

BJT characteristic curves

- Input characteristics (base curve): I_B versus V_{BE} (for specified values of V_{CE}).

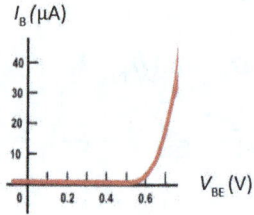

Figure 4.25   Input characteristics

- Output characteristics (collector curve): I_C versus V_{CE} (for specified values of I_B).

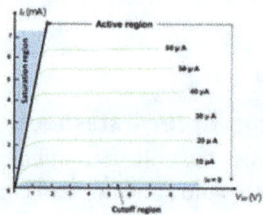

Figure 4.26   Output characteristics

BJT – switch and amplifier

- The transistors have two basic functions: "switching" (switch) or "amplification" (amplifier).
- The transistors can operate in three different regions: the cutoff region, saturation region, and active region.
- The BJT as amplifiers and switches
  - A transistor in the active (linear) region can work as an amplifier.
  - A transistor in the saturation or cutoff regions can work as a switch.

Table 4.7   BJT regions of operation

Region	Bias	Condition	I_C	Acts as
Cutoff	C–B: RB; B–E: RB	$V_{BE} < V_{RB},\ I_B \approx 0,\ I_C \approx 0$ $(I_B = I_{BCO},\ I_C = I_{CEO})$	$I_C \approx 0$	Open switch (off)
Saturation	C–B: FB; B–E: FB	$V_{CE} < V_{BR}:\ I_B \uparrow \rightarrow I_C \uparrow$ $V_{CE} > V_{BR}:\ I_B \uparrow \rightarrow \overrightarrow{I_C}$	$I_C = I_{C(sat)}$	Closed switch (on)
Active	C–B: RB; B–E: FB	$I_B \uparrow \rightarrow I_C \uparrow$	$I_C = \beta_{DC} I_B$	Amplifier

Transistor DC load line analysis

DC load line: a line drawn on the collector characteristic curve by connecting the cutoff and saturation points.

- Saturation point: $V_{CE} = V_{CE(sat)},\ I_C = I_{C(max)}$
- Cutoff point: $V_{CE} = V_{CE(max)} \approx V_{CC},\ I_C \approx 0$
- The region between cutoff and saturation is known as an active region.

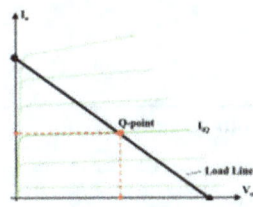

Figure 4.27   DC load line

BJT as an amplifier

- Amplifier: an electronic device that produces an output that is larger than the input. It increases the amplitude of signals.

Table 4.8   Voltage and current notation (in this book)

Quantity	Letter	Subscript	Example
DC	Capital	Capital	I_B, I_C, V_{CE}
AC	Lowercase	Capital (except r_e)	i_B, i_C, v_{CE}, r_e
Instantaneous	Lowercase	Lowercase	i_c, i_e, v_c, v_{ce}

BJT AC analysis

- Transistor modeling: use an equivalent circuit (model) to represent the AC characteristics of the transistor.
- AC equivalent circuit: an AC equivalent circuit can be made by reducing all DC sources to zero (short circuit all DC voltage sources).

The r_e transistor model

- r_e model: an AC equivalent circuit that can be used to predict the performance of a BJT.
- r_e: the internal AC resistance looking into the emitter terminal of a BJT.
- Calculating r_e:

$$r_e = \frac{v_B}{i_E} \qquad v_B = v_{in} - i_B R_B, \quad i_E \approx i_C, \quad i_C = \frac{v_C}{R_C}$$

Common emitter voltage gain

- Voltage gain (A_V): $A_V = \dfrac{\text{Output voltage}}{\text{Input voltage}} = \dfrac{v_{out}}{v_{in}}$

- The voltage gain in the common emitter configuration:

$$A_V = \frac{v_{out}}{v_{in}} \approx \frac{R_C}{r_e}$$

Table 4.9   The BJT as a switch

BJT	Switch	Voltage and current	Minimum/maximum current
Cutoff	Switch off (nonconducting)	$V_{in} = 0$, $I_B \approx 0$, $I_C \approx 0$ $I_B = I_{BCO}$, $I_C = I_{CEO}$	Minimum current $I_C \approx 0$
Saturation	Switch on (conducting)	$V_{in} \neq 0$ ($V_{in} \leq V_{BR}$) $V_{CE} < V_{BR}$: $I_B \uparrow \rightarrow I_C \uparrow$ $V_{CE} = V_{BR}$: $I_B \uparrow \rightarrow \overline{I_C}$	Maximum current $I_C = I_{C(sat)}$

Self-test

4.1 1. A BJT is a (   )-controlled transistor that uses both electrons and holes charge carriers to conduct current.

2. A PNP transistor is a transistor in which one N-type semiconductor layer is placed between two P-type semiconductor layers, and (   ) are the majority charge carriers.

3. Beta is the ratio of the emitter current to the (   ) current.

4. Hybrid parameter (h_{FE}) is the same as the (   ) but is more widely used in transistor datasheets.

5. A transistor operates with a base current of 5 mA and an emitter current of 300 mA. Determine the DC current gains beta (β_{DC}) and alpha (α_{DC}).

6. A transistor operates with a base current of 80 A and a collector current of 7 mA. Determine the DC current gains beta (β_{DC}) and alpha (α_{DC}).

7. If the junction temperature increases (the collector current I_C does not change), the DC current gain β_{DC} will (   ).

4.2 8. A common emitter circuit has an R_B of 15 kΩ, R_C of 130 Ω, supply voltage V_{BB} of 5 V, V_{CC} of 15 V, and β_{DC} of 150. Determine the DC currents I_B, I_C, I_E, and the DC voltages V_{CE} and V_{CB}. (Assuming a silicon transistor)

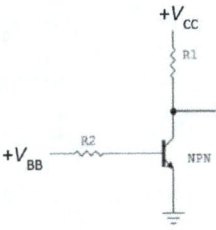

Figure 4.28   Ch 4 – No. 8, self-test

9. Output characteristics is the graph of the (   ) current versus the collector-emitter voltage.

10. The transistors have two basic functions: "(   )" or "amplification."

11. The transistors can operate in three different regions: the cutoff region, saturation region, and (   ) region.

12. Saturation region: both B–E junction and C–B junction are forward biased, and the transistor acts as a (   ) switch.

13. A DC load line is a line drawn on the collector characteristic curve by connecting the cutoff and (   ) points.

14. The DC load line is the locus of all possible operating points at which BJT remains in the (   ) region.

15. Determine if the BJT in Figure 4.29 is saturated for the following values: $R_B = 25$ kΩ, $R_C = 4$ kΩ, $V_{BB} = 5$V, $V_{CC} = 15$V, $\beta_{DC} = 75$, and $V_{CE(max)} = 0.2$ V. (Assuming a silicon transistor)

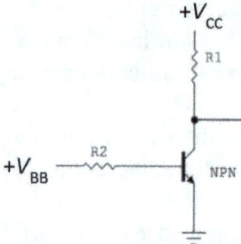

Figure 4.29   Ch 4 – No. 15, self-test

4.3 16. An AC equivalent circuit can be made by reducing all DC sources to (   ).
17. r_e is the (   ) AC resistance looking into the emitter terminal of a BJT.
18. A common emitter circuit has r_e of 25 Ω, a 20 mV input signal at the base of the transistor, and an output voltage of 1.5 V at the collector. Determine the collector resistance R_C.

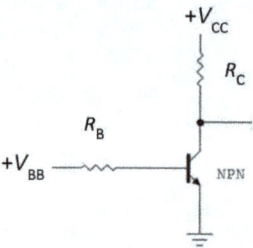

Figure 4.30   Ch 4 – No. 18, self-test

19. What the minimum value of input voltage is required to saturate the BJT in Figure 4.31, if the BJT has the following values: R_1 is 2 kΩ, R_2 is 60 kΩ, V_{BB} is 6 V, V_{CC} is 15 V, and β_{DC} is 100? (Assuming a silicon transistor)

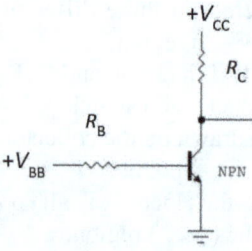

Figure 4.31   Ch 4 – No. 19, self-test

Chapter 5

DC biasing of BJTs

Chapter outline

5.1   Introduction to DC biasing

5.1.1   DC load line and operation point

DC biasing

- DC biasing: establishing a predetermined level of DC voltages or currents in a transistor amplifier.
- Recall – amplifier: an electronic device that produces an output that is larger than the input. It increases the amplitude of signals (voltage, current, power, etc.).
- The purpose of DC biasing a transistor: a DC bias current or voltage is added (superpositioned) to the AC signal so that the signal is proper amplified.
    - DC biasing a BJT can set up the initial DC values (operating point).
    - DC biasing a BJT can ensure the BJT operate in the active (linear) region (i.e., not in cutoff or saturation modes).
      For amplification, the transistor must operate in the active or linear region.

Linear versus nonlinear amplifier

- Linear amplifier: the output signal is directly proportional to the input signal.
  An amplifier is used to increase the amplitude of an AC signal without altering its shape →
  a linear amplifier.

- Nonlinear amplifier: the output signal is not directly proportional to the input signal.

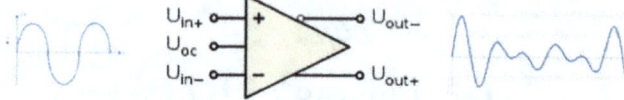

Figure 5.1   Nonlinear amplifier

The operating point (Q-point)

- The Q-point (quiescent point or bias point): the DC voltage or current at a specific point within the operation characteristic curve of a circuit with no AC input signal applied.

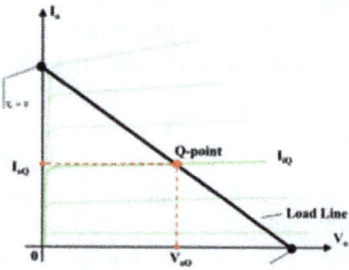

Figure 5.2   Q-point

- The Q-point (I_C and V_{CE}) of a BJT circuit is obtained from the values of the collector current I_C versus collector-emitter voltages V_{CE} (for a specific I_B on the DC load line).

Bias a transistor   →   establish a Q-point

DC load line and Q-point

- Different operating points: the Q-point could be at any of the intersection points between the DC load line and the output characteristic curves.
- The Q-point moves along the DC load line:
  - V_{CE} decreases when I_B and I_C increase (if V_{BB} increases).
    $I_C = \beta_{DC} I_B, \; V_{CE} = V_{CC} - I_C R_C$
    The Q-point moves along the load line from the lower Q-point (Q_1) to Q_2 or the higher Q point (Q_3).

$$V_{BB} \uparrow \rightarrow I_B \uparrow \rightarrow I_C \uparrow \rightarrow V_{CE} \downarrow \qquad (Q_1 \rightarrow Q_2 \rightarrow Q_3)$$

  - V_{CE} increases when I_B and I_C decrease (if V_{BB} decreases). The Q-point moves along the load line from the higher Q-point (Q_3) to Q_2 or the lower Q-point (Q_1).

$$V_{BB} \downarrow \rightarrow I_B \downarrow \rightarrow I_C \downarrow \rightarrow V_{CE} \uparrow \qquad (Q_3 \rightarrow Q_2 \rightarrow Q_1)$$

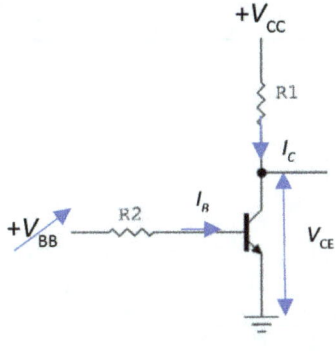

Figure 5.3(a) Circuit values affect the Q-point

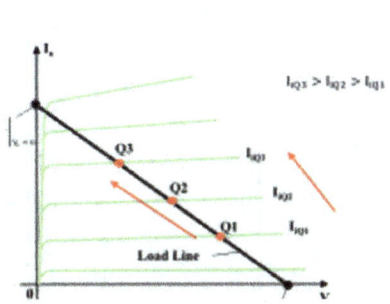

Figure 5.3(b) Q-point moves

Saturation and cutoff points

- Actual/ideal saturation point:

  - Actual saturation point: $V_{CE} = V_{CE\,(sat)}$, $I_C = \dfrac{V_{CC} - V_{CE(sat)}}{R_C}$ $V_{CE\,(sat)} = V_{BR}$

  - Ideal saturation point: $V_{CE} = 0$, $I_C = \dfrac{V_{CC}}{R_C}$ Ignore $V_{CE\,(sat)}$

- Actual/ideal cutoff points:

  - Actual cutoff point: $V_{CE} = V_{CE\,(max)} - I_{CBO} R_C$, $I_C = I_{CBO}$

    $V_{CE} = V_{CC} - I_C R_C$

  - Ideal cutoff point: $V_{CE} = V_{CC}$, $I_C = I_B = 0$ Ignore I_{CBO}

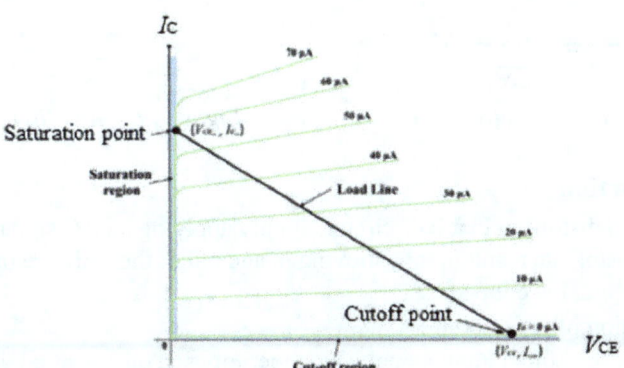

Figure 5.4 Saturation and cutoff points

Table 5.1   Saturation and cutoff points

	Ideal		Actual	
Cutoff (off)	V_{CE}:	$V_{CE} = V_{CC}$	V_{CE}:	$V_{CE} = V_{CE\,(max)} - I_{CBO} R_C$
	I_C:	$I_C = 0$	I_C:	$I_C = I_{CBO}$
Saturation (on)	V_{CE}:	$V_{CE} = 0$	V_{CE}:	$V_{CE} = V_{ce(sat)}$
	I_C:	$I_C = \dfrac{V_{CC}}{R_C}$	$V_{CE\,(sat)} = V_{BR}$, $V_{BR} = 0.7\text{V}$ for Si or 0.3V for Ge	
			I_C:	$I_C = \dfrac{V_{CC} - V_{CE(sat)}}{R_C}$

5.1.2   Midpoint biasing and active (linear) operation

Active (linear) operation

- Active (linear) region: the region between cutoff and saturation along the DC load line.
  - Below saturation
  - Above cutoff
- BJT amplifier: a transistor in the active or linear region can work as an amplifier to amplify a small AC input signal.

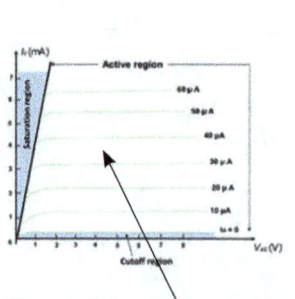

Figure 5.5   Active region

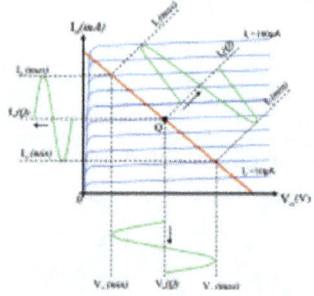

Figure 5.6   Amplifying

Signal distortion

- Waveform distortion in a BJT circuit: simply applying an AC signal to the base of a transistor can result in it by moving in and out of the active region of operation (the signal is clipped).
- The position of the Q-point:
  Assume a sinusoidal input signal v_{in} is superimposed on V_{BB} of a BJT amplifier.
  - If the Q-point is above the center, the output waveform will be limited by saturation (the positive peak will be clipped).

- If the Q-point is below the center, the output waveform will be limited by cutoff (the negative peak will be clipped).
- A large input signal: if the input signal is too large, the output waveform will be limited by both saturation and cutoff (both positive and negative peaks will be clipped).

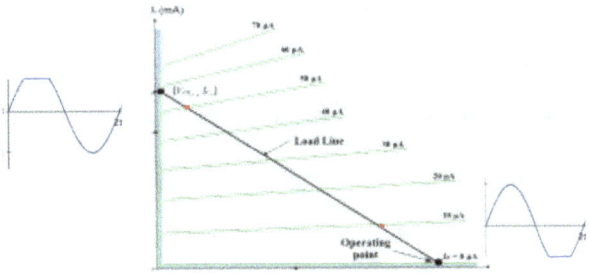

Figure 5.7   Signal distortion

Amplifier midpoint biasing

- Midpoint biasing: an amplifier with a centered Q-point on the DC load line. It represents the most efficient use of the amplifiers range for operation with AC input signals.
- Midpoint biasing can provide amplification of an AC input signal without distortion or clipping to the output waveform.
  The complete cycle of amplified signal will appear at the output of the circuit.

Example: Calculate the values of I_C and V_{CE} in Figure 5.8. Determine whether the circuit is midpoint biased. Given $R_B = 25$ kΩ, $R_C = 200$ Ω, $V_{BB} = 8$ V, $V_{CC} = 12$ V, and $\beta_{DC} = 100$.   (Assuming a silicon BJT)

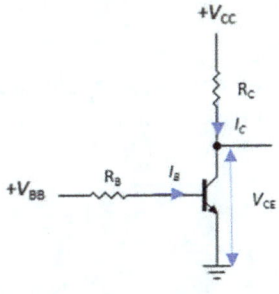

Figure 5.8   Midpoint biased?

- Given: $R_B = 25$ kΩ, $R_C = 200$ Ω, $V_{BB} = 8$ V, $V_{CC} = 12$ V, $\beta_{DC} = 100$
- Find: I_C and V_{CE}. Midpoint biased? (the Q-point centered on the DC load line?)

- Solution:

$$I_B: \quad I_B = \frac{V_{BB} - V_{BE}}{R_B} = \frac{8V - 0.7\ V}{25\ k\Omega} \approx 0.29\ mA$$

$$I_C: \quad I_C = \beta_{DC}\, I_B = (100)\,(0.29\ mA) = 29\ mA$$

$$V_{CE}: \quad V_{CE} = V_{CC} - I_C\, R_C = 12\ V - (29\ mA)\,(0.2\ k\Omega)$$

Milli: 10^{-3}; Kilo: 10^{3}

$$= 12\ V - 5.8\ V \approx 6.2\ V$$

The Q-point is near the center of the DC load line.

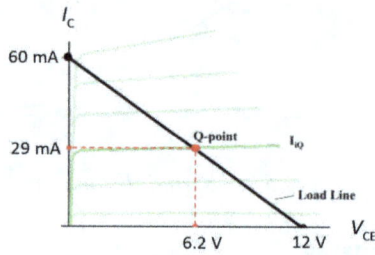

Figure 5.9   The Q-point is near the center

5.1.3   *BJT biasing and stability*

The goal of transistor biasing

- Recall – biasing: establishing a predetermined level of DC voltages or currents in a transistor amplifier so that the AC signal is proper amplified.
- Transistors must be properly biased to establish a known Q-point to operate correctly and produce the desired amplification effect.
- Correct DC biasing of the BJT also establishes its initial AC operating region with an undistorted amplified output signal.

Types of transistor biasing

- The commonly used methods of transistor biasing from one source of supply (V_{CC}):
  - Base bias (fixed bias or base resistor method)

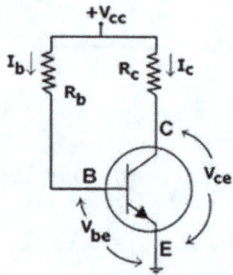

Figure 5.10   Base bias

– Voltage-divider bias

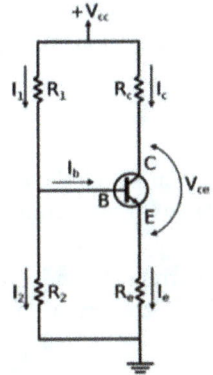

Figure 5.11 Voltage-divider bias

– Collector-feedback bias

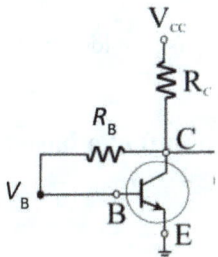

Figure 5.12 Collector feedback

– Emitter bias

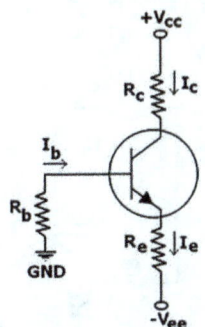

Figure 5.13 Emitter bias

It uses two supply voltages (V_{CC} and V_{EE}).

- It is desirable that transistor circuits have a single source of supply (simplicity and economy).
- Most of the above methods have the same principle of obtaining the required value of I_B and I_C from a single-voltage source V_{CC} with the zero AC input signals.

Stability

- Amplifier stability: a condition in which the currents and voltages remain relatively constant over a wide range of temperatures and transistor parameter beta β (transistor current gain).
- Q-point stability: collector current I_C and collector-emitter voltage V_{CE} are almost independent of transistor temperature variations in β value.

5.2   Methods of BJT biasing

5.2.1   Base bias

Base bias (fixed base or base resistor method)
- Base bias circuit:
- The required base current of I_B is provided by V_{CC} which flows through the base resistor R_B.

The DC operating point (Q-point) of a base bias circuit

- Base current: $I_B = \dfrac{V_{cc} - V_{BE}}{R_B}$

  Derive: $V_{cc} = I_B R_B + V_{BE}$
  KVL (Kirchhoff's voltage law): $\sum V = \sum E$ (from the base-emitter loop)
- Collector current: $I_C = \beta_{DC} I_B$
- Collector-emitter voltages: $V_{CE} = V_{CC} - I_C R_C$

  Derive: $V_{cc} = I_C R_C + V_{CE}$ KVL (from the collector-emitter loop)

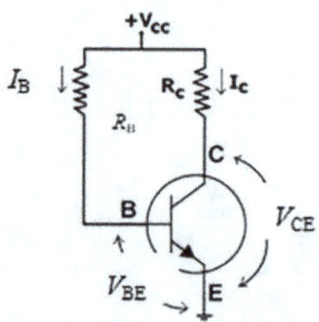

Figure 5.14   Base bias circuit

Effect of transistor parameter β on the Q-point

- $I_C = \beta_{DC}\, I_B = \beta_{DC}\, \dfrac{V_{CC} - V_{BE}}{R_B}$ $I_B = \dfrac{V_{cc} - V_{BE}}{R_B}$

- The V_{CC}, V_{BE}, and R_B are fixed known quantities.
  $V_{BE} = V_{BR} = 0.7$ V for Si or 0.3 V for Ge

- The I_C for the base bias circuit depends on the value of β_{DC} for a transistor. The value of β_{DC} changes with temperature.

Stability of a base bias circuit

- Temperature $\uparrow \to \beta_{DC} \uparrow \to I_C \uparrow \to V_{CE} \downarrow \to$ Q-point shift $\to$ unstable

  $I_C = \beta_{DC}\, I_B$, $V_{CE}\downarrow = V_{CC} - I_C\uparrow R_C$

  Temperature $\downarrow \to \beta_{DC} \downarrow \to I_C \downarrow \to V_{CE} \uparrow \to$ Q-point shift $\to$ unstable

  $V_{CE}\uparrow = V_{CC} - I_C\downarrow R_C$

  Temperature $\uparrow \to I_{CBO}\uparrow \to I_B \uparrow \to$ Q-point shift $\to$ unstable

  I_{CBO} – reverse leakage current

- The stability of the base bias circuit is poor, it is not a desirable configuration for linear operation because its Q-point varies with β_{DC} value.
- The base bias method is generally used in switching circuits.

5.2.2   *Voltage-divider bias*

Voltage-divider bias circuit:

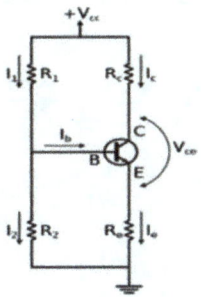

Figure 5.15   Voltage-divider bias circuit

Stability of a voltage-divider bias circuit

- Voltage-divider bias is one of the most frequently used biasing circuits.
- It uses a voltage divider circuit to provide good Q-point stability (I_C and V_{CE} are almost independent of transistor parameter β_{DC}).
  The two resistors R_1 and R_2 form a voltage divider as the name implies – voltage-divider bias.
- Voltage-divider configuration makes the BJT circuit independent of temperature changes in β_{DC} value.

Calculating DC voltages and currents

- The input resistance in the B–E loop ($R_{IN(base)}$):

$$R_{IN(base)} \approx \beta_{DC} R_E$$

Derive: $R_{IN(base)} = \dfrac{V_{BE} + I_E R_E}{I_B} \approx \dfrac{I_E R_E}{I_B}$ $V_{BE} \approx V_{BR} \approx 0$

$$\approx \dfrac{\beta_{DC} \not{I_B} R_E}{\not{I_B}} \approx \beta_{DC} R_E$$ $I_E \approx I_C = \beta_{DC} I_B$

When the emitter resistor R_E is viewed from the base, it appears to be larger than its actual value because of the DC current gain β_{DC}.

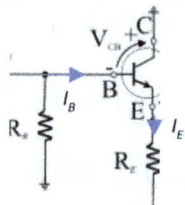

Figure 5.16 $R_{IN(base)}$

- Collector current I_C – precision approach:

$$I_C \approx \dfrac{V_B - V_{BE}}{R_E}, \qquad V_B = V_{CC} \dfrac{R_2 \, // \, R_{IN(base)}}{R_1 + R_2 \, // \, R_{IN(base)}}$$

Derive: $I_C \approx I_E$ ← $I_E = \dfrac{V_E}{R_E}$ ← $V_E = V_B - V_{BE}$

$$V_B = V_{CC} \dfrac{R_2 \, // \, R_{IN(base)}}{R_1 + R_2 \, // \, R_{IN(base)}} \qquad V_2 = E \dfrac{R_2}{R_1 + R_2}, \quad R_{IN(base)} \approx \beta_{DC} R_E, \quad R_1 \backslash\backslash R_2 = \dfrac{R_1 R_2}{R_1 + R_2}$$

$$I_C \approx I_E = \dfrac{V_E}{R_E} = \dfrac{V_B - V_{BE}}{R_E} \qquad V_B = V_{CC} \dfrac{R_2 \backslash\backslash R_{IN(base)}}{R_1 + R_2 \backslash\backslash R_{IN(base)}}$$

Figure 5.17   Derive V_B

- Collector current I_C – approximate approach:

$$I_C \approx \dfrac{V_B - V_{BE}}{R_E}, \quad V_B \approx V_{CC} \dfrac{R_2}{R_1 + R_2}, \quad \text{if } R_{IN(base)} >> R_2 \text{ (check if } \beta_{DC} R_E > 10 \, R_2)$$

Derive: $I_C \overset{?}{\approx} I_E \quad \leftarrow \quad I_E = \dfrac{V_E^?}{R_E} \quad \leftarrow \quad V_E \overset{?}{=} V_B - V_{BE}$

$I_C \approx I_E = \dfrac{V_E}{R_E} = \dfrac{V_B - V_{BE}}{R_E}$

$V_B = V_{CC} \dfrac{R_2 \mathbin{//} R_{IN(base)}}{R_1 + R_2 \mathbin{//} R_{IN(base)}}$
$\qquad\qquad\qquad\qquad R_{IN(base)} \approx \beta_{DC} R_E$

If $R_{IN(base)} \gg R_2$ (check if $\beta_{DC} R_E > 10\, R_2$): $V_B \approx V_{CC} \dfrac{R_2}{R_1 + R_2}$
$\qquad R_2 \parallel R_{IN(base)} \approx R_2$

$I_C \approx \dfrac{V_B - V_{BE}}{R_E}, \quad V_B \approx V_{CC} \dfrac{R_2}{R_1 + R_2}$

(Example: $R_2 = 1\,\Omega$, $R_{IN(base)} = 100\,\Omega$)

$R_2 \parallel R_{IN(base)} = \dfrac{R_2\, R_{IN(base)}}{R_2 + R_{IN(base)}} = \dfrac{(1)(100)}{1 + 100} \approx 0.99\,\Omega \approx 1\,\Omega = R_2)$
$\qquad R_1 \parallel R_2 = \dfrac{R_1\, R_2}{R_1 + R_2}$

- Collector-emitter voltage V_{CE}:

$V_{CE} = V_{CC} - I_C R_C - I_E R_E$
$\qquad\qquad\qquad\qquad$ KVL: $V_{CC} = I_C R_C + V_{CE} + I_E R_E$

$\quad\;\; = V_{CC} - I_C (R_C + R_E)$
$\qquad\qquad\qquad\qquad I_E \approx I_C$

Stability of a voltage-divider bias circuit

- The voltage-divider bias is a very stable circuit. The Q-point values of I_C and V_{CE} are almost independent of variations in the current gain β_{DC} value, and the BJT always remains in the active-linear region.

- The voltage-divider bias is commonly used in the design of BJT amplifier circuits.

Example: Determine the Q-point for the circuit shown in Figure 5.15, if R_1 is 15 kΩ, R_2 is 4.7 kΩ, R_C is 2 kΩ, R_E is 1.5 kΩ, V_{CC} is 12V, and β_{DC} is 100. (Assuming a silicon BJT)

- Given: $R_1 = 15$ kΩ, $R_2 = 4.7$ kΩ, $R_C = 2$ kΩ, $R_E = 1.5$ kΩ, $V_{CC} = 12$V, and $\beta_{DC} = 100$.
- Find: Q-point (values of I_C and V_{CE})
- Solution:
  - $\boldsymbol{R_{IN(base)}}$: Thinking process: $R_{IN(base)} \overset{?}{\gg} R_2 \;\rightarrow\; R_{IN(base)} \approx \beta\, DC\, R\,E \;\rightarrow\; \beta_{DC} R_E \overset{?}{>} 10\, R_2$

$R_{IN(base)} \approx \beta_{DC} R_E = (100)(1.5\text{ k}\Omega) = 150\text{ k}\Omega$

$150\text{ k}\Omega \overset{?}{>} 10\,(4.7)\text{ k}\Omega = 47\text{ k}\Omega, \quad \text{yes!} \qquad \beta_{DC} R_E \overset{?}{>} 10\, R_2$

It can use the approximate approach.

- I_C: Thinking process: $I_C \approx I_E$ $\rightarrow$ $I_E = \dfrac{V_E}{R_E}$ $\rightarrow$ $V_E = V_B - V_{BE}$

$$V_B \approx V_{CC}\frac{R_2}{R_1 + R_2} = 12\ \text{V}\frac{4.7\ \text{k}\Omega}{15\ \text{k}\Omega + 4.7\ \text{k}\Omega} \approx 2.86\ \text{V}$$

$$I_C \approx I_E = \frac{V_E}{R_E} = \frac{V_B - V_{BE}}{R_E} = \frac{2.86\ \text{V} - 0.7\ \text{V}}{1.5\ \text{k}\Omega} \approx 1.44\ \text{mA} \quad \text{Kilo: } 10^3;\ \text{Milli: } 10^{-3}$$

V_{CE}: $V_{CE} = V_{CC} - I_C(R_C + R_E)$

$\qquad = 12\ \text{V} - (1.44\ \text{mA})(2\ \text{k}\Omega + 1.5\ \text{k}\Omega) = 6.96\ \text{V}$

5.2.3   Emitter bias

Emitter bias circuit

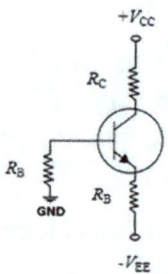

Figure 5.18   *Emitter bias*

- Emitter bias configuration uses two supply voltages, one positive $(+V_{CC})$ and one negative $(-V_{EE})$, which are equal but opposite in polarity.
- $(-V_{EE})$ forward biases the base-emitter junction while $(+V_{CC})$ reverse biases the collector-base junction.

Emitter bias circuit analysis

- Collector current I_C: $I_C \approx I_E = \dfrac{-V_{EE} - V_{BE}}{\dfrac{R_B}{\beta_{DC}} + R_E}$

Derive: $I_B R_B + V_{BE} + I_E R_E = -V_{EE}$ $\qquad\qquad$ KVL

$\qquad \dfrac{I_E}{\beta_{DC}}R_B + V_{BE} + I_E R_E = -V_{EE}$ $\qquad\qquad$ $I_C = \beta\,DCI_B,\ I_B \approx \dfrac{I_E}{\beta_{DC}}$

$\qquad I_E\left(\dfrac{R_B}{\beta_{DC}} + R_E\right) + V_{BE} = -V_{EE}$ $\qquad\qquad$ Factor out I_E

$$I_E \approx I_C = \frac{-V_{EE} - V_{BE}}{\dfrac{R_B}{\beta_{DC}} + R_E}$$

$$I_C \approx I_E$$

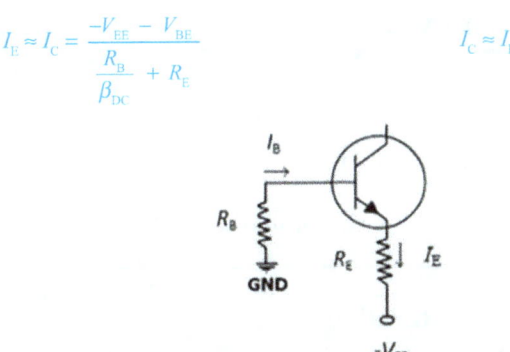

Figure 5.19 Derive I_C

- Collector-emitter voltages V_{CE} : $V_{CE} = V_{CC} - V_{EE} - I_C(R_C + R_E)$

Derive: $I_C R_C + V_{CE} + I_E R_E = V_{CC} - V_{EE}$ KVL

$V_{CE} + I_C(R_C + R_E) = V_{CC} - V_{EE}$ $I_E \approx I_C$

$V_{CE} = V_{CC} - V_{EE} - I_C(R_C + R_E)$

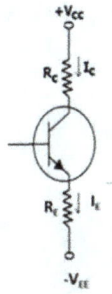

Figure 5.20 Derive V_{CE}

Example: Determine the Q-point for the circuit shown in Figure 5.18, if R_B is 56 kΩ, R_C is 510 Ω, R_E is 1 kΩ, V_{CC} is 10 V, V_{EE} is –10 V, and β_{DC} is 200. (Assuming a silicon BJT)

- Given: $R_B = 56$ kΩ, $R_C = 510$ Ω, $R_E = 1$ kΩ, $V_{CC} = 10$ V, $V_{EE} = -10$ V, and $\beta_{DC} = 200$.
- Find: Q-point (values of I_C and V_{CE})
- Solution:

$$I_C : I_C \approx I_E = \frac{-V_{EE} - V_{BE}}{\dfrac{R_B}{\beta_{DC}} + R_E}$$

$$I_C = \frac{-(-10\ V) - 0.7\ V}{\dfrac{56\ k\Omega}{200} + 1\ k\Omega} \approx \mathbf{7.27\ mA} \qquad V_{BE} \approx V_{BR};\ \text{Kilo: } 10^3;\ \text{Milli: } 10^{-3}$$

V_{CE}: $V_{CE} = V_{CC} - V_{EE} - I_C(R_C + R_E)$

$$= 10\ V - (-10\ V) - (7.27\ mA)(0.51\ k\Omega + 1\ k\Omega) \approx \mathbf{9.02\ V}$$

Emitter bias – cutoff and saturation

- Saturation (when $V_{CE} = 0$): $I_{C\,(sat)} = \dfrac{V_{CC} - V_{EE}}{R_E + R_C}$

  Derive: $V_{CE} = V_{CC} - V_{EE} - I_C(R_C + R_E)$

  When $V_{CE} = 0$: $0 = V_{CC} - V_{EE} - I_{C(sat)}(R_C + R_E)$

  Solve for I_C : $I_{C\,(sat)} = \dfrac{V_{CC} - V_{EE}}{R_E + R_C}$

- Cutoff (when $I_C = 0$): $V_{CE(cutoff)} = V_{CC} - V_{EE}$

  Derive: $V_{CE} = V_{CC} - V_{EE} - I_C(R_C + R_E)$

  When $I_C = 0$: $V_{CE} = V_{CC} - V_{EE}$

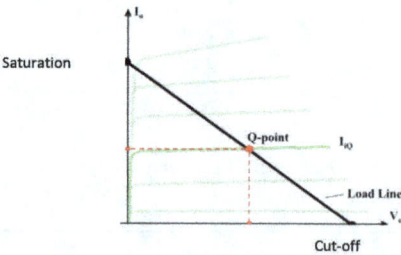

Figure 5.21   Cutoff and saturation

Example: Determine the saturation and cutoff values of $I_{C(sat)}$ and $V_{CE(cutoff)}$ for the circuit shown in Figure 5.18, if R_B is 56 kΩ, R_C is 510 Ω, R_E is 1 kΩ, V_{CC} is 10V, V_{EE} is −10V, and β_{DC} is 200.

- Given: R_B = 56 kΩ, R_C = 510 Ω, R_E = 1 kΩ, V_{CC} = 10 V, V_{EE} = −10 V, and β_{DC} = 200.

- Find: $I_{C(sat)}$ and $V_{CE(cutoff)}$.

- Solution:

$$- \quad I_{C(sat)} \colon I_{C\ (sat)} = \frac{V_{CC} - V_{EE}}{R_E + R_C} = \frac{10\ V - (-10\ V)}{1\ k\Omega + 0.51\ k\Omega} \approx 13.25\ mA \quad \text{Kilo: } 10^3;\ \text{Milli: } 10^{-3}$$

$$- \quad V_{CE(cutoff)} \colon V_{CE(cutoff)} = V_{CC} - V_{EE} = 10\ V - (-10\ V) = 20\ V$$

DC load line of Figure 5.19 (for the above two examples)

- Q-point: $I_C = 7.27\ mA$, $V_{CE} = 9.02\ V$

- Saturation point: $V_{CE(sat)} \approx 0\ V$, $I_{C(sat)} = 13.25\ mA$

- Cutoff point: $V_{CE(cutoff)} = 20\ V$, $I_{C(sat)} \approx 0\ mA$

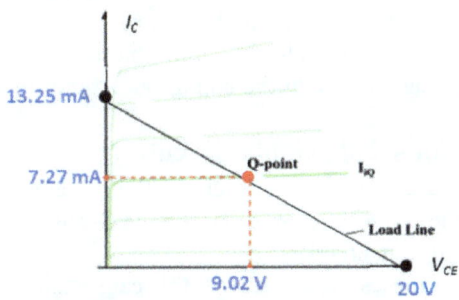

Figure 5.22   DC load line

- The Q-point of this emitter bias circuit is near the middle of its linear operating range that is approximately halfway between cutoff and saturation.
- The BJT can remain in the active-linear region if the design (choice of resistors) is well done.

Stability of an emitter bias circuit

- Recall: $I_C \approx I_E = \dfrac{-V_{EE} - V_{BE}}{\dfrac{R_B}{\beta_{DC}} + R_E}$, $\qquad V_{CE} = V_{CC} - V_{EE} - I_C (R_C + R_E)$

  If $R_E \gg \dfrac{R_B}{\beta_{DC}}$ $\qquad I_C \approx I_E \approx \dfrac{-V_{EE} - V_{BE}}{R_E}$

  If $V_{EE} \gg V_{BE}$: $\qquad I_C \approx I_E \approx \dfrac{-V_{EE}}{R_E}$

- Emitter bias is a stable circuit: change in the transistor DC current gain (β_{DC}) value nearly does not affect the Q-point values of I_C and V_{CE}.
- Emitter bias circuit provides good Q-point stability. The Q-point is almost independent of variation in temperature (T) or the DC current gain (β_{DC}).

5.2.4   Collector-feedback bias

Collector-feedback bias circuit

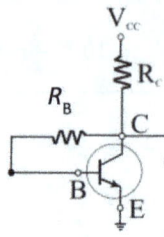

Figure 5.23 Collector-feedback bias

- The base resistor R_B of the collector-feedback bias circuit is connected to the collector C rather than to V_{CC} as in the figure above.
- Negative feedback: a type of self-regulation that tends to reduce a process by applying the output against the initial condition. It tends to stabilize the system.

Stability of a collector-feedback bias circuit
- This collector-to-base negative-feedback configuration ensures that the BJT is always biased in the active-linear region, and the Q-point is only slightly dependent on the DC current gain (β_{DC}).
- If temperature T increases, β_{DC} increases, this causes the collector current I_C to increase.

$$T\uparrow \;\;\to\;\; \beta_{DC}\uparrow \;\;\to\;\; I_C\uparrow \qquad\qquad I_C = \beta_{DC} I_B$$

- If I_C increases, the voltage drop across R_C increases, thereby causing V_C to decrease. When V_C decreases, there is a decrease in voltage across R_B, which decreases I_B fed back to the base. This, in turn, decreases the I_C, correcting the original increase in V_C.

$$I_C\uparrow \;\to I_C R_C\uparrow \;\to\; V_C\downarrow \to I_B R_B\downarrow \to I_B\downarrow \qquad V_C\downarrow = V_{CC} - I_C R_C\uparrow$$

$$I_C\downarrow \longleftarrow \qquad\qquad\qquad\qquad V_C\downarrow = V_{BE} + (I_B R_B)\downarrow$$

Collector-feedback circuit analysis

- Collector current I_C : $I_C \approx I_E = \dfrac{V_{CC} - V_{BE}}{R_C + \dfrac{R_B}{\beta_{DC}}}$

Derive: $I_C R_C + I_B R_B + V_{BE} = V_{CC}$ $\qquad\qquad\qquad$ KVL

$$I_C R_C + \frac{I_C}{\beta_{DC}} R_B = V_{CC} - V_{BE} \qquad\qquad I_C = \beta_{DC} I_B , \; I_B = \frac{I_C}{\beta_{DC}}$$

$$I_C \left(R_C + \frac{R_B}{\beta_{DC}} \right) = V_{CC} - V_{BE}$$

$$I_C = \frac{V_{CC} - V_{BE}}{R_C + \dfrac{R_B}{\beta_{DC}}}$$

- Collector-emitter voltages V_{CE} : $V_{CE} \approx V_C = V_{CC} - I_C R_C$

Example: Determine values of I_C and V_{CE} for the circuit in Figure 5.24, if R_B is 150 kΩ, R_C is 5 kΩ, V_{CC} is 12 V, and β_{DC} is 100. (Assuming a silicon BJT)

- Given: R_B = 150 kΩ, R_C = 5 kΩ, V_{CC} = 12 V, and β_{DC} = 100.
- Find: I_C and V_{CE}
- Solution:

$$I_C: \; I_C \approx I_E = \frac{V_{CC} - V_{BE}}{R_C + \dfrac{R_B}{\beta_{DC}}} = \frac{12\,V - 0.7\,V}{5\,k\Omega + \dfrac{150\ k\Omega}{100}} = 1.74\ mA \qquad \text{Kilo: } 10^3;\ \text{Milli: } 10^{-3}$$

$$V_{CE}: \; V_{CE} \approx V_C = V_{CC} - I_C R_C = 12\ V - (1.74\ mA)\,(5\ k\Omega) = 3.3\ V$$

Summary

Q-point
- The Q-point (quiescent point or bias point): the DC voltage or current at a specific point within the operation characteristic curve of a circuit with no AC input signal applied.

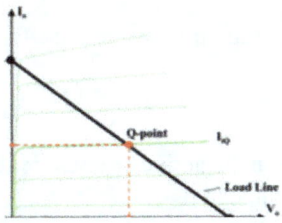

Figure 5.24   The Q-point

- The Q-point (I_C and V_{CE}) of a BJT circuit is obtained from the values of the collector current I_C versus collector-emitter voltages V_{CE} (for a specific I_B on the DC load line).

    Bias a transistor   →   establish a Q-point

DC load line and Q-point

- The Q-point moves along the DC load line:

$$V_{BB} \uparrow \rightarrow I_B \uparrow \rightarrow I_C \uparrow \rightarrow V_{CE} \downarrow \qquad (Q_1 \rightarrow Q_2 \rightarrow Q_3)$$

$$V_{BB} \downarrow \rightarrow I_B \downarrow \rightarrow I_C \downarrow \rightarrow V_{CE} \uparrow \qquad (Q_3 \rightarrow Q_2 \rightarrow Q_1)$$

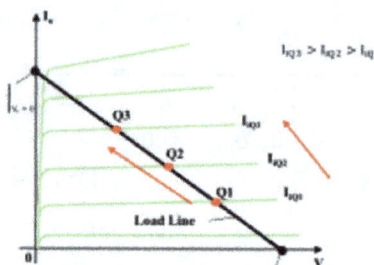

Figure 5.25   Q-point moves

Table 5.2   Actual/ideal saturation and cutoff points

	Ideal	Actual
Cutoff (off)	V_{CE}: $V_{CE} = V_{CC}$ I_C: $I_C = 0$	V_{CE}: $V_{CE} = V_{CE\,(max)} - I_{CBO}\, R_C$ I_C: $I_C = I_{CBO}$
Saturation (on)	V_{CE}: $V_{CE} = 0$ I_C: $I_C = \dfrac{V_{CC}}{R_C}$	V_{CE}: $V_{CE} = V_{CE(sat)}$ $\qquad V_{CE(sat)} = V_{BR}$ I_C: $I_C = \dfrac{V_{CC} - V_{CE(sat)}}{R_C}$

Active (linear) operation

- Active (linear) region: the region between cutoff and saturation along the DC load line.
  - Below saturation
  - Above cutoff
- BJT amplifier: a transistor in the active or linear region can work as an amplifier to amplify a small AC input signal.

Signal distortion

- Waveform distortion in a BJT circuit: simply apply an AC signal to the base of a transistor can result in it by moving in and out of the active region of operation (the signal is clipped).
- A large input signal: if the input signal is too large, the output waveform will be limited by both saturation and cutoff (both positive and negative peaks will be clipped).

Amplifier midpoint biasing

- Midpoint biasing: an amplifier with a centered Q-point on the DC load line. It represents the most efficient use of the amplifiers range for operation with AC input signals.
- Midpoint biasing can provide amplification of an AC input signal without distortion or clipping to the output waveform.
- The commonly used methods of transistor biasing from one source of supply (V_{CC}):
  - Base bias (fixed bias or base resistor method)
  - Voltage-divider bias
  - Collector-feedback bias
  - Emitter bias (uses two supply voltage V_{CC} and V_{EE}).

Stability

- Amplifier stability: a condition in which the currents and voltages remain relatively constant over a wide range of temperatures and transistor parameter beta β (transistor current gain).
- Q-point stability: collector current I_C and collector-emitter voltage V_{CE} are almost independent of transistor temperature T variations in β value.

Table 5.3 Methods of transistor biasing

Biasing	Base bias	Voltage-divider bias	Emitter bias	Collector-feedback bias
Circuit				
I_C	$I_C = \beta_{DC} I_B$	$I_C = \dfrac{V_B - V_{BE}}{R_E}$ $V_B = V_{CC} \dfrac{R_2 \,//\, R_{IN(base)}}{R_1 + R_2 \,//\, R_{IN(base)}}$ If $R_{IN(base)} >> R_2 : V_B \approx V_{CC} \dfrac{R_2}{R_1 + R_2}$ $R_{IN(base)} \approx \beta_{DC}\, R_E$	$I_C \approx I_E = \dfrac{-V_{EE} - V_{BE}}{\dfrac{R_B}{\beta_{DC}} + R_E}$	$I_C \approx I_E = \dfrac{V_{CC} - V_{BE}}{R_C + \dfrac{R_B}{\beta_{DC}}}$
V_{CE}	$V_{CE} = V_{CC} - I_C R_C$	$V_{CE} = V_{CC} - I_C (R_C + R_E)$	$V_{CE} = V_{CC} - V_{EE} - I_C (R_C + R_E)$	$V_{CE} \approx V_C = V_{CC} - I_C R_C$
Stability	Unstable (It is used in switching circuits.)	Stable (It is commonly used in the design of BJT amplifier circuits.)	Stable (It uses two supply voltages V_{CC} and V_{EE}.)	Stable (It uses a negative-feedback configuration.)

Self-test

5.1 1. A bipolar junction transistor is a (   )-controlled transistor that uses both electrons and holes charge carriers to conduct current.

2. The purpose of DC biasing a transistor: a DC bias current or voltage is added to the AC signal so that the signal is proper (   ).

3. A transistor in the active or (   ) region can work as an amplifier to amplify a small AC input signal.

4. If the Q-point is below the center, the output waveform will be limited by (   ).

5. Midpoint biasing can provide amplification of an AC input signal without (   ) or clipping to the output waveform.

6. Calculate the values of I_C and V_{CE} in Figure 5.26. Determine whether the circuit is midpoint biased. Given R_B is 30 kΩ, R_C is 300 Ω, V_{BB} is 5 V, V_{CC} is 15 V, $R_B = 30$ kΩ, $R_C = 300$ Ω, $V_{BB} = 5$ V, $V_{CC} = 15$ V, and β_{DC} is 100. (Assuming a silicon BJT)

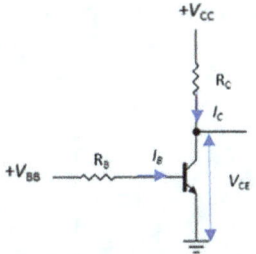

Figure 5.26 Ch 5: No. 6, self-test

5.2 7. Voltage-divider bias is one of the most frequently used (   ) circuits.

8. Voltage-divider configuration makes the BJT circuit independent of (   ) changes in β_{DC} value.

9. Determine the Q-point for the circuit shown in Figure 5.27, if R_1 is 20 kΩ, R_2 is 15 kΩ, R_C is 3 kΩ, R_E is 2 kΩ, V_{CC} is 15 V, and β_{DC} is 150. (Assuming a silicon BJT)

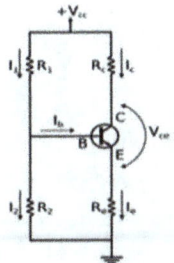

Figure 5.27   Ch 5: No. 9, self-test

10. Determine the Q-point for the circuit shown in Figure 5.28, if $R_B = 60$ kΩ, $R_C = 550$ Ω, $R_E = 1.5$ kΩ, $V_{CC} = 15$ V, $V_{EE} = -15$ V, β_{DC} is 150. (Assuming a silicon BJT)

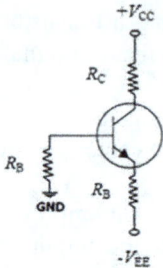

Figure 5.28   Ch 5: No.10, self-test

11. Determine the saturation and cutoff values of $I_{C(sat)}$ and $V_{CE(cutoff)}$ for the circuit shown in Figure 5.29, if R_B is 45 kΩ, R_C is 400 Ω, R_E is 2 kΩ, V_{CC} is 15V, V_{EE} is −15V, and β_{DC} is 150. (Assuming a silicon BJT)

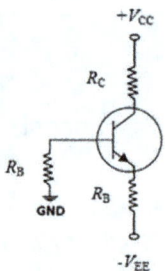

Figure 5.29   Ch 5: No. 11, self-test

12. Emitter bias circuit provides good Q-point stability. The Q-point is almost independent of variation in (   ) or the DC current gain β.

13. Determine values of I_C and V_{CE} for the circuit shown in Figure 5.30, if R_B is 200 kΩ, R_C is 10 kΩ, V_{CC} is 15V, β_{DC} is 150, and β_{DC} is 150. (Assuming a silicon BJT)

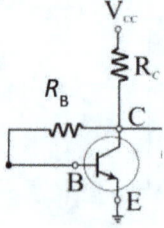

Figure 5.30   Ch 5: No. 13, self-test

Chapter 6

AC analysis of BJT circuits – BJT amplifiers

6.1 Transistor amplifiers

6.1.1 Capacitors in transistor amplifier

Amplifier

- Recall – amplifier: an electronic device that increases the voltage, current, or power of a input signal. It is a circuit that has a power gain greater than one.
- Amplifier gain: the ratio of output to input in the form of power, voltage, and current. It is a measure of the ability of an amplifier.

  - Power gain: $A_\text{P} = \dfrac{P_\text{out}}{P_\text{in}}$

  - Voltage gain: $A_\text{V} = \dfrac{V_\text{out}}{V_\text{in}}$

  - Current gain: $A_\text{i} = \dfrac{i_\text{out}}{i_\text{in}}$

- BJT amplifier: a BJT in the linear region around an operating point can work as an amplifier to amplify a small AC input signal (low-level signal such as wireless signal).

A small-signal BJT amplifier with voltage-divider bias

- Circuit of a BJT amplifier with voltage-divider bias:

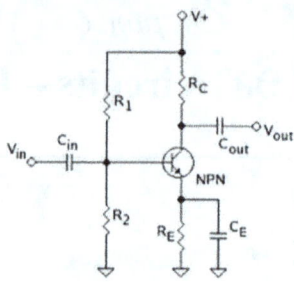

Figure 6.1   A BJT amplifier with voltage-divider bias

- Recall – the purpose of biasing is to stabilize the Q-Point (DC operation) in the desired region of operation in response to an AC input signal.
- A small AC voltage (V_{in}) is applied to the input and passed to output (V_{out}) through coupling capacitors (C_{in} and C_{out}) that are used to pass an AC signal without disturbing the Q-point of the circuit.
- A bypass capacitor (C_E) on the emitter resistor (R_E) to bypass desirable signals to the ground and to increase gain.

Capacitor review

- Capacitor for AC: a capacitor acts as a short circuit in AC (high frequencies). Capacitor for DC: a capacitor acts as an open circuit in DC (low frequencies).
- Recall: the capacitive reactance is proportional to the inverse of the frequency $(X_C = \dfrac{1}{2\pi fC})$

AC: $f \uparrow\uparrow \;\to\; X_C \downarrow\downarrow \;\to\; 0$ $\qquad\qquad X_C \downarrow\downarrow = \dfrac{1}{2\pi f \uparrow\uparrow C}$

At higher and higher frequencies, X_C approaches zero – short circuit (pass AC).

DC: $\qquad f \downarrow\downarrow \to X_C \uparrow\uparrow \to \infty$ $\qquad\qquad X_C \uparrow\uparrow = \dfrac{1}{2\pi f \downarrow\downarrow C}$

At lower and lower frequencies, X_C approaches infinity – open circuit (block DC).

- A capacitor blocks DC, so it can be used to pass an AC signal (e.g., audio) without disturb the Q-point of the circuit.
- Capacitor: pass AC signal (short circuit)
              block DC (open circuit)

The function of capacitors in BJT amplifier

- Coupling capacitors (C_{in} and C_{out})
  - Capacitive coupling (AC coupling): the transfer of AC signals or electrical energy from one segment of a circuit to another using a capacitor.

- Coupling capacitor: a capacitor that is connected between two nodes of a circuit such that only the AC signal can pass through while DC is blocked (a DC blocking capacitor).
- A coupling capacitor can prevent the internal source resistance (R_s), and the load resistance (R_L), from changing DC bias or Q-point.
- Bypassing capacitor (C_E)
  - Bypass capacitor: a capacitor that is used to short AC signal to the ground so that the AC noise is removed. It also acts as an open circuit for a DC and maintains the DC bias (not affecting DC operation).
  - A bypass capacitor can eliminate the effect of voltage spikes that may be on an AC signal and reduce the power supply noise.

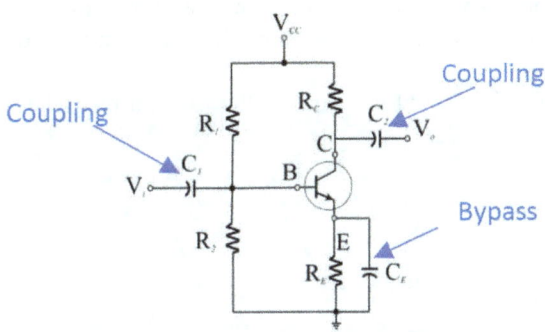

Figure 6.2   Capacitors in BJT amplifier

6.1.2   Superposition of DC and AC

Voltage and current notation (in this book)

- DC quantities use capital letters with capital subscripts.

  Example: I_B, I_C, I_E, V_{CE}, V_C, V_{BE},   ...

- AC quantities use lowercase letters with capital subscripts (except r_e).

  Example: i_B, i_C, i_E, v_{CE}, v_{BE}, v_{CB}, r_e,   ...

- AC RMS (root mean square) quantities use uppercase letters with lowercase subscripts.

  Example: I_c, I_e, V_c, V_{ce}, R_e,   ...

- Instantaneous quantities use lowercase letters with lowercase subscripts.

  Example: i_c, i_e, v_c, v_{ce},   ...

Table 6.1   Voltage and current notation

Quantity	Letter	Subscript	Example
DC	capital	capital	I_B, I_C, V_{CE}
AC	lowercase	capital	i_B, i_C, v_{CE}, r_e
AC RMS	capital	lowercase	I_c, I_e, V_c, V_{ce}, R_e
Instantaneous	lowercase	lowercase	i_c, i_e, v_c, v_{ce}

Superposition of DC and AC waveforms

- Superposition: two or more waves combine in a manner which is the alge-braic sum of the waveforms produced by each independent wave acting separately.
- Superposition of AC and DC: combine the waveforms of the AC and DC.

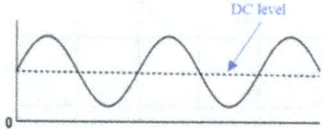

Figure 6.3   Superposition of AC and DC

Example: DC: an amplifier has a constant DC operating point (Q-point).
AC: the AC signal is usually what will be amplified.

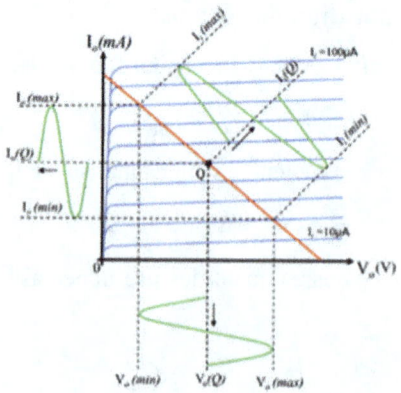

Figure 6.4   Amplifying

6.1.3   *Transistor r parameters equivalent model*

Transistor AC model

- Model: an informative representation of an object, device, system, etc. that shows what it looks like or how it works and predicts the behavior of a device in a particular operating condition.
- Transistor model: an informative representation of circuit elements that best shows the actual behavior of a transistor under specific operating regions.
- A transistor AC model can be represented by *r*-parameters or *h*-parameters equivalent circuit. The circuit uses internal parameters to represent the BJT's operation.
- *r*-parameter or *h*-parameters model: an equivalent circuit that can be used to accurately predict the performance of a BJT. It can be used to quickly estimate the input impedance, gain, and operating conditions of BJT amplifiers.

Transistor *r*-parameters equivalent circuit

- Derive transistor *r*-parameter AC equivalent circuit: in any transistor, $I_c = \beta_{dc} I_b$, thus $\beta_{dc} I_b$ can be thought of as a current generator.
- The *r* parameter AC equivalent circuit is shown below:

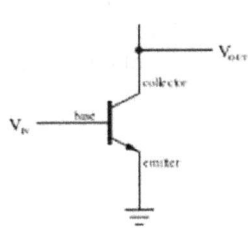

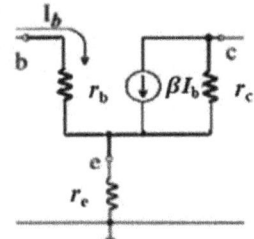

Figure 6.5(a)   BJT

Figure 6.5(b)   r parameter AC equivalent circuit

AC current gains of a transistor

- AC current gain beta (β_{ac}): the ratio of the AC collector current I_c to the AC base current I_b.

$$\beta_{ac} = \frac{I_c}{I_b}$$

- AC current gain alpha (α_{ac}): the ratio of the AC collector current I_c to the AC emitter current I_e.

$$\alpha_{ac} = \frac{I_c}{I_e}$$

Internal resistance of a transistor

- BJT internal AC resistance r: a small AC resistance looking into the terminal of a transistor. It is like the dynamic resistance for a forward-biased diode.
- r is equivalent to the slope of voltage-current of the PN diode.

Example: $r_e = \dfrac{\Delta V_{be}}{\Delta I_e}$

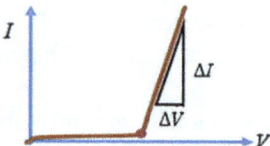

Figure 6.6   V-I of the PN diode

- AC base resistance r_b: it represents the base region AC resistance that depends upon the doping density. r_b is usually very small (the base is thin and lightly doped) and can be ignored in calculations (it can be replaced by a short circuit).
- AC collector resistance r_c: it represents reverse-biased collector-base junction resistance. it is typically much larger (reverse-biased) than R_c and can be neglected in calculations (it can be replaced by an open circuit).
- AC emitter resistance r_e: it represents the forward-biased base-emitter junction resistance and has a low value. At room temperature, the following equations can be used to find an approximate value for r_e.

$$r_e = \frac{26 \text{ mV}}{I_E \text{ (mA)}}$$ At room temperature

Simplified transistor *r* parameter equivalent circuit

- r_b: very small – short circuit
- r_c: very large – open circuit
- The collector branch acts as a current source ($\beta_{ac} I_b$).

Figure 6.7   Simplified BJT r parameter equivalent circuit

Table 6.2   r parameters and current gains

r parameter	Formula
AC emitter resistance r_e	$r_e = \dfrac{26 \text{ mV}}{I_E \text{ (mA)}}$ (at room temperature)
AC base resistance r_b	very small – short circuit
AC collector resistance r_c	very large – open circuit
AC current gain β_{ac}	$\beta_{ac} = \dfrac{I_c}{I_b}$
AC current gain α_{ac}	$\alpha_{ac} = \dfrac{I_c}{I_e}$

6.1.4   Transistor h parameters equivalent model

Transistor h parameters

- Hybrid parameters or *h* parameters: every linear circuit (or network) having input and output terminals can be analyzed by four parameters called hybrid (*h*) parameters.

Figure 6.8   Network

- Four general *h* parameters:
  - Input resistance or impedance h_i: the ratio of the input voltage V_i to the input current I_i (with the output short-circuited).

  $$h_i = \frac{V_i}{I_i}$$ Subscript "i" – input

  - Output conductance or admittance h_o: the ratio of the output current I_o to the output voltage V_o (with the input open-circuited).

  $$h_o = \frac{I_o}{V_o}$$ Subscript "o" – output

  - Forward current gain h_f: the ratio of the load current I_o to the input current I_i (with the output short-circuited).

  $$h_f = \frac{I_o}{I_i}$$ Subscript "f" – forward bias

  - Reverse voltage gain h_r: the ratio of the input voltage V_i to the output voltage V_o (with the input open-circuited).

  $$h_r = \frac{V_i}{V_o}$$ Subscript "r" – reverse bias

- Four hybrid parameters measured in ohm, mho (or siemens), or dimensionless. Hybrid means "mixed". Since there is a mixture of units involved, they are called hybrid parameters.
- Hybrid parameters h_f and h_r: they are the same as the r parameters (AC current gain β_{ac} or α_{ac}) but are more widely used in transistor datasheets (they can easily be measured).
- Small-signal analysis: both r and h parameters are valid only for small-signal analysis (the amplifier's linear region of operation).

  The BJT is a non-linear device so only for a small signal can use small-signal approximation and treat the BJT as a linear device.

Table 6.3   h *parameters*

h parameters	Subscript	Condition	Formula	Unit
Input impedance or resistance h_i	i – input	Output short circuit	$h_i = \dfrac{V_i}{I_i}$	ohm
Output admittance or conductance h_o	o – output	Input open circuit	$h_o = \dfrac{I_o}{V_o}$	mho
Forward current gain h_f	f – forward	Output short circuit	$h_f = \dfrac{I_o}{I_i}$	dimensionless
Reverse voltage gain h_r	r – reverse	Input open circuit	$h_r = \dfrac{V_i}{V_o}$	dimensionless

Introduction to fundamental BJT amplifier circuit configurations

- There are three fundamental configurations for a transistor amplifier:
  - Common-emitter amplifier
  - Common-collector amplifier
  - Common-base amplifier
- Common-emitter amplifier (CE):
  - The input signal is applied between the base and emitter terminals of the BJT.
  - The output is taken between the collector and emitter terminals.
  - The emitter is a common connection terminal for both input and output.

    Emitter – common; V_{in} – base; V_{out} – collector

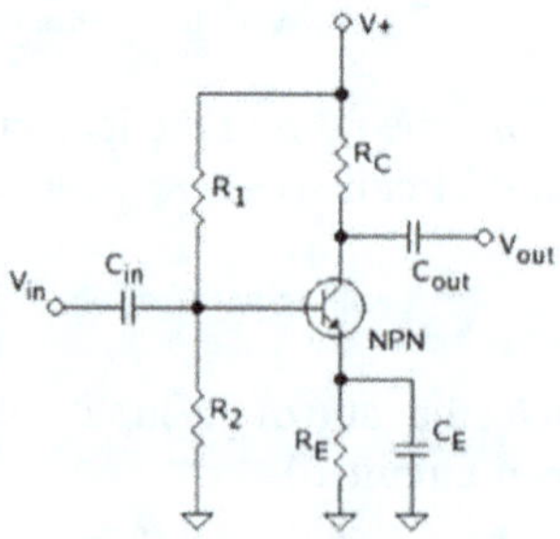

Figure 6.9   Common-emitter amplifier

- Common-collector amplifier (CC):
  - The input signal is applied between the base and emitter terminals.
  - The output is taken from the emitter branch of the BJT.
  - The collector terminal is a common connection point for both input and output. (The collector terminal is "grounded" through the power supply. Both the signal source and the output share the collector lead as a common connection point.)

    Collector – common; V_{in} – base; V_{out} – emitter

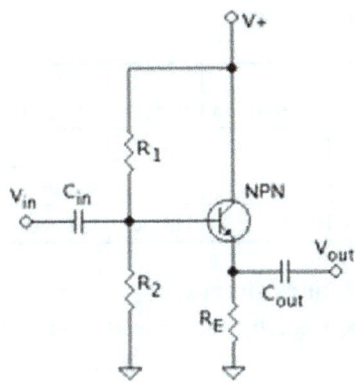

Figure 6.10   Common-collector amplifier

- Common-base amplifier (CB):
  - The input signal is applied to the emitter branch of the BJT.
  - The output is taken from the collector branch of the BJT.
  - The base terminal is a common connection point for both input and output.

    Base – common; V_{in} – emitter; V_{out} – collector

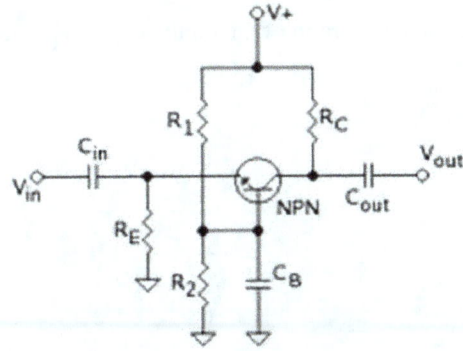

Figure 6.11   Common-base amplifier

6.1.5   h *parameters for three configurations of the transistors*

h **parameters for three BJT amplifier configurations**

- Subscript letter: the second subscript letter is added to the four *h* parameters to define different types of amplifier configurations.
- Common-emitter amplifier (CE) – add a subscript e.   e – emitter
- Common-collector amplifier (CC) – add a subscript c.   c – collector
- Common-base amplifier (CB) – add a subscript b.   b – base

Table 6.4   Types of amplifier configurations

Amplifier	Abbreviation	Input	Output	Common terminal	Subscript
Common-emitter amplifier	CE	base	collector	emitter	e
Common-collector amplifier	CC	base	emitter	collector	c
Common-base amplifier	CB	emitter	collector	base	b

Common-emitter (CE) h **parameters**

- Input resistance or impedance h_{ie}: the ratio of the input voltage V_b to the input current I_b.

$$h_{ie} = \frac{V_i}{I_i} = \frac{V_b}{I_b}$$

- Output conductance or admittance h_{oe}: the ratio of the output current I_c to the output voltage V_{ce}.

$$h_{oe} = \frac{I_o}{V_o} = \frac{I_c}{V_{ce}}$$

- Forward current gain h_{fe}: the ratio of the output current I_c to the input current I_b.

$$h_{fe} = \frac{I_o}{I_i} = \frac{I_c}{I_b}$$

- Reverse voltage gain h_{re}: the ratio of the input voltage V_b to the output voltage V_c.

$$h_{re} = \frac{V_o}{V_i} = \frac{V_b}{V_c}$$

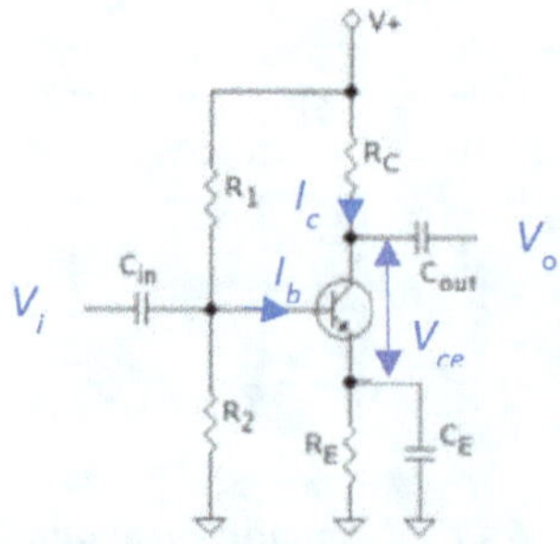

Figure 6.12   CE amplifier

Common-collector (CC) *h* parameters

- Input resistance (or impedance) h_{ic}: the ratio of the input voltage V_b to the input current I_b.

$$h_{ic} = \frac{V_i}{I_I} = \frac{V_b}{I_b}$$

- Output conductance (or admittance) h_{oc}: the ratio of the output current I_e to the output voltage V_e.

$$h_{oc} = \frac{I_o}{V_o} = \frac{I_e}{V_e}$$

- Forward current gain h_{fc}: the ratio of the output current I_e to the input current I_b.

$$h_{fc} = \frac{I_o}{I_I} = \frac{I_e}{I_b}$$

- Reverse voltage gain h_{rc}: the ratio of the input voltage V_b to the output voltage V_e.

$$h_{rc} = \frac{V_o}{V_i} = \frac{V_b}{V_e}$$

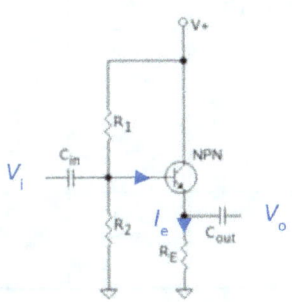

Figure 6.13   CC amplifier

Common-base (CB) amplifier *h* parameters

- Input resistance (or impedance) h_{ib}: the ratio of the input voltage V_e to the input current I_e.

$$h_{ib} = \frac{V_i}{I_i} = \frac{V_e}{I_e}$$

- Output conductance (or admittance) h_{ob}: the ratio of the output current I_c to the output voltage V_c.

$$h_{ob} = \frac{I_o}{V_o} = \frac{I_c}{V_c}$$

- Forward current gain h_{fb}: the ratio of the output current I_c to the input current I_e.

$$h_{fb} = \frac{I_o}{I_i} = \frac{I_c}{I_e}$$

- Reverse voltage gain h_{rb}: the ratio of the input voltage V_e to the output voltage V_c:

$$h_{rb} = \frac{V_o}{V_i} = \frac{V_e}{V_c}$$

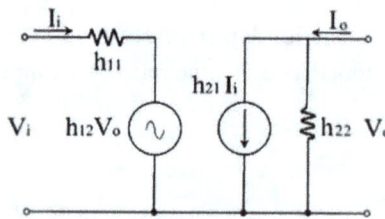

Figure 6.14   CB amplifier

Table 6.5   h *parameters for three configurations*

Configuration	Subscript	h_i	h_o	h_f	h_r
Common-emitter (CE)	e	$h_{ie} = \dfrac{V_b}{I_b}$	$h_{oe} = \dfrac{I_c}{V_{ce}}$	$h_{fe} = \dfrac{I_c}{I_b}$	$h_{re} = \dfrac{V_b}{V_c}$
Common-collector (CC)	c	$h_{ic} = \dfrac{V_b}{I_b}$	$h_{oc} = \dfrac{I_e}{V_e}$	$h_{fc} = \dfrac{I_e}{I_b}$	$h_{rc} = \dfrac{V_b}{V_e}$
Common-base (CB)	b	$h_{ib} = \dfrac{V_e}{I_e}$	$h_{ob} = \dfrac{I_c}{V_c}$	$h_{fb} = \dfrac{I_c}{I_e}$	$h_{rb} = \dfrac{V_e}{V_c}$

h parameters main equivalent circuit

Figure 6.15   h parameter main equivalent circuit

(Subscripts 11, 12, 21, and 22 could be ie, oe, fe, ...)

Conversion between h and r parameters

- *h* parameters and *r* parameters are similar, designed for small signals, involve using an equivalent circuit to model transistor behavior, but *h* parameters are more widely used in transistor datasheets (datasheets provide *h* parameters).

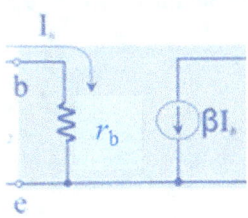
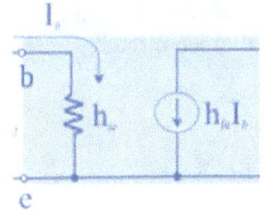

Figure 6.16(a)   r parameters equivalent

Figure 6.16(b)   h parameters equivalent

- The forward current gain h_{fe} is the same as the β_{ac} that can be seen from h and r parameters common-emitter equivalent circuits.

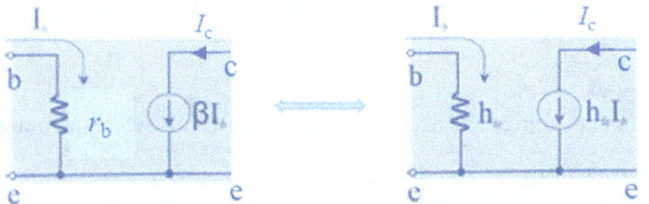

Figure 6.17   h and r conversion

$$h_{fe} = \beta_{ac} = \frac{I_c}{I_b}$$

6.2   Transistor small-signal analysis

6.2.1   *Transistor common-emitter amplifier*

DC analysis
- Recall: common-emitter (CE) circuit
  - Emitter – common
  - V_{in} – base
  - V_{out} – collector

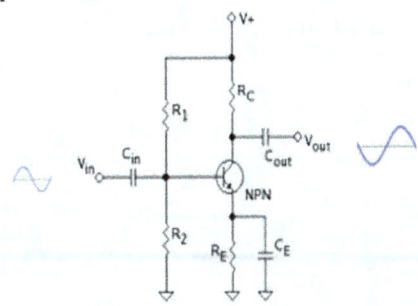

Figure 6.18   Common-emitter amplifier

- The DC equivalent circuit of a CE amplifier is obtained by replacing all capacitors by an open circuit.

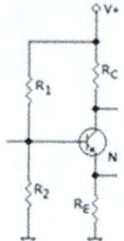

Figure 6.19   The DC equivalent circuit of a CE amplifier

- The capacitive reactance is proportional to the inverse of the frequency $(X_C = \dfrac{1}{2\pi fC})$.

- Recall: a capacitor acts as an open circuit in DC (low frequencies).

$$\text{DC: } f\downarrow\downarrow \rightarrow X_C \uparrow\uparrow \rightarrow \infty \rightarrow \text{ open circuit} \qquad X_C = \frac{1}{2\pi fC}$$

- Q-point of a common-emitter circuit:

$$I_C \approx I_E = \frac{V_E}{R_E} = \frac{V_B - V_{BE}}{R_E} \qquad V_B = V_{CC}\frac{R_2 \,//\, R_{IN(base)}}{R_1 + R_2 \,//\, R_{IN(base)}}, \; R_{IN(base)} \approx \beta_{DC}\, R_E$$

$$\text{If } R_{IN(base)} \gg R_2, \qquad V_B \approx V_{CC}\frac{R_2}{R_1 + R_2}$$

$$V_{CE} \approx V_C - V_E = I_C R_C - (V_B - V_{BE}) \qquad V_C = I_C R_C, \quad V_E = V_B - V_{BE}$$

$$I_B = \frac{I_C}{\beta_{DC}}$$

AC equivalent circuit

- The AC equivalent circuit of a CE amplifier is obtained by replacing all capacitors by a short circuit and setting DC voltage sources to zero and replacing them by ground.

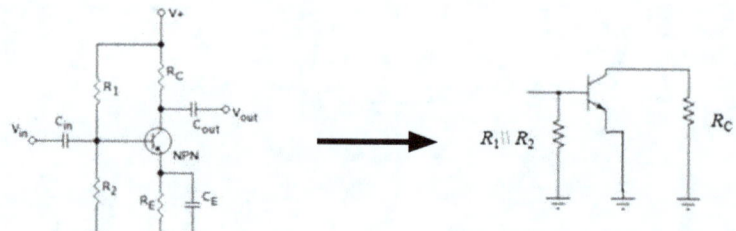

Figure 6.20(a)   A CE amplifier

Figure 6.20(b)   AC equivalent circuit of a CE amplifier

- Capacitors are replaced by short circuits in AC analysis
  - Recall: a capacitor acts as a short circuit in AC (high frequencies).
  - AC: $f\uparrow\uparrow \rightarrow X_C \downarrow\downarrow \rightarrow 0 \rightarrow$ short circuit $X_C = \dfrac{1}{2\pi f C}$
- DC source is replaced by a ground in AC analysis
  - DC source does not generate any changes, therefore from an AC analysis point of view, their contribution is zero.
  - For an AC signal, a zero voltage DC source is the same as ground.

Input resistance

- R_{ib}: the input resistance at the base of the transistor.

$$R_{ib} = \frac{V_b}{I_b} = \frac{I_e r_e}{\dfrac{I_e}{\beta_{ac}}} = \beta_{ac} r_e \qquad\qquad V_b = I_e r_e, \quad I_b = \frac{I_e}{\beta_{ac}}, \quad h_{fe} = \beta_{ac}$$

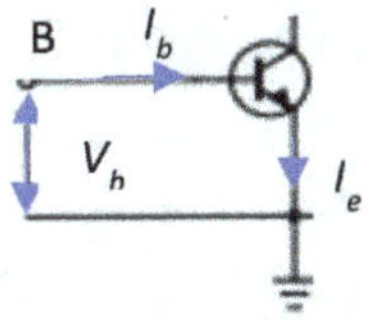

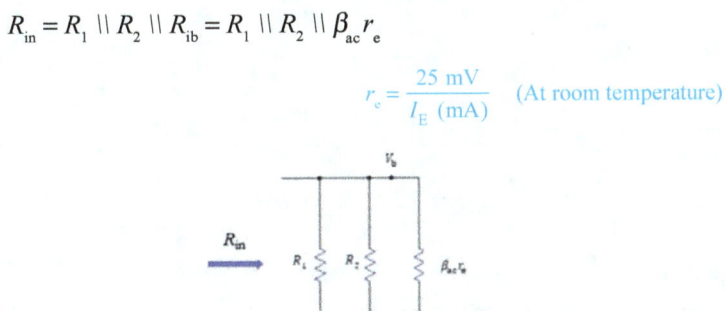

Figure 6.21   Input resistance at the base

- R_{in}: the total circuit input resistance presented to the AC source.

$$R_{in} = R_1 \parallel R_2 \parallel R_{ib} = R_1 \parallel R_2 \parallel \beta_{ac} r_e$$

$$r_e = \frac{25\ \text{mV}}{I_E\ (\text{mA})} \qquad (\text{At room temperature})$$

Figure 6.22   Total circuit input resistance

Output resistance

- R_o: the output resistance presented to the load of the circuit.
  It acts as the source resistance for its load.

$$R_o = R_C \parallel r_c$$

- r_c: the AC collector resistance r_c represents reverse-biased collector-base junction resistance.
- r_c is typically much larger than the collector resistor R_C and can be replaced by an open circuit.

$$R_o \approx R_C \qquad\qquad \text{If } r_c \gg R_C$$

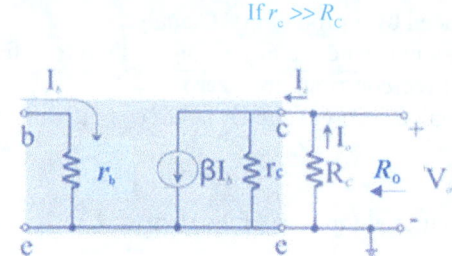

Figure 6.23   *Output resistance*

The AC voltage V_b at the base

- If R_S (the source resistance) $= 0$: $V_b = V_{in}$

- If $R_S \neq 0$: $$V_b = V_{in} \frac{R_{in}}{R_S + R_{in}} \qquad R_{in} = R_1 /\!/ R_2 /\!/ \beta_{ac} r_e$$

- If $R_S \ll R_{in}$: $$V_b \approx V_{in} \qquad V_b = V_{in} \frac{R_{in}}{R_S + R_{in}}$$

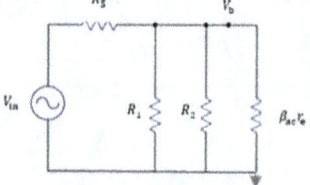

Figure 6.24   **V**$_b$ *at the base*

Voltage gain

- One of the key aspects of the amplifiers and their circuit design is the voltage gain. It is a measure of the ability of an amplifier.
- Amplifier voltage gain A_V: the ratio of output to input in the form of voltage.
  A_V is how much the output voltage is greater than the input voltage.
- Voltage gain without the load resistor R_L:

$$A_V = \frac{V_{out}}{V_{in}} = \frac{V_c}{V_b} = \frac{I_c R_C}{I_e r_e} \approx \frac{I_e R_C}{I_e r_e} = \frac{R_C}{r_e} \qquad I_c \approx I_e$$

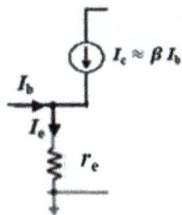

Figure 6.25   Derive the voltage gain

- Voltage gain with the load resistor R_L:

$$A_V = \frac{V_{out}}{V_{in}} = \frac{R_C \backslash\backslash R_L}{r_e}$$

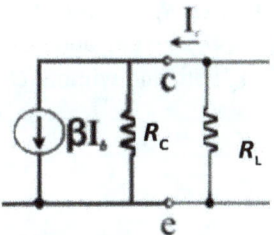

Figure 6.26   Voltage gain with R_L

A bypass capacitor can increase voltage gain
- Recall: a capacitor acts as a short circuit in AC.
- An amplifier without a bypass capacitor: the voltage gain for an amplifier without a bypass capacitor – R_E is in series with r_e:

$$A_V = \frac{R_C}{r_e + R_E}$$

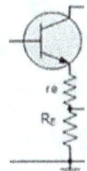

Figure 6.27   An amp without a bypass C

- An amplifier with a bypass capacitor: when a bypass capacitor C_E is connected with an emitter resistance R_E, the voltage gain of C_E amplifier increases.

$$A_V = \frac{R_C}{r_e}$$ C_E is short-circuited.

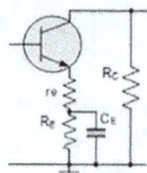

Figure 6.28   An amp with a bypass C

Example: For the BJT amplifier of Figure 6.18, determine the Q-point, input and output resistance R_{in} and R_{out}, AC signal voltage at the base V_b, and voltage gain A_V both with and without the bypass capacitor C_E and load R_L, if R_1 is 15 kΩ, R_2 is 4.7 kΩ, R_C is 2 kΩ, R_E is 1.5 kΩ, R_S is 100 Ω, R_L is 47 kΩ, V_{CC} is 12V, V_{in} is 8 mV, β_{DC} is 100, and h_{fe} is 200. (Assuming a silicon BJT)

- Given: R_1 = 15 kΩ, R_2 = 4.7 kΩ, R_C = 2 kΩ, R_E = 1.5 kΩ, R_S = 100 Ω, R_L = 47 kΩ, V_{CC} = 12V, V_{in} = 8 mV, β_{DC} = 100, and h_{fe} = 200.
- Find: Q-point, R_{in}, R_o, V_b, A_V with and without C_E, and A_V with and without R_L (with C_E).
- Solution:

DC (Q-point): I_C and V_{CE}

- I_C: Thinking process: $I_C \approx I_E = \dfrac{V_E}{R_E} \overset{?}{\rightarrow} V_E = V_B \overset{?}{-} V_{BE} \rightarrow$ if $R_{IN(base)} \overset{?}{\gg} R_2$

$$\rightarrow V_B \approx V_{CC} \frac{R_2}{R_1 + R_2}$$

$$R_{IN\,(base)} = \beta_{DC} R_E = (100)\,(1.5\ \text{k}\Omega) = 150\ \text{k}\Omega \qquad R_{IN(base)} \gg R_2$$

$$V_B \approx V_{CC} \frac{R_2}{R_1 + R_2} = 12\ \text{V}\frac{4.7\ \text{k}\Omega}{15\ \text{k}\Omega + 4.7\ \text{k}\Omega} \approx 2.86\ \text{V}$$

$$V_E = V_B - V_{BE} = 2.86\ \text{V} - 0.7\ \text{V} = 2.16\ \text{V}$$

$$I_C \approx I_E = \frac{V_E}{R_E} = \frac{2.16\ \text{V}}{1.5\ \text{k}\Omega} = 1.44\ \text{mA} \qquad \text{Milli: } 10^{-3};\ \text{Kilo: } 10^3$$

- V_{CE}: Thinking process: $V_{CE} = V_C \overset{?}{-} V_E \rightarrow V_C = V_{CC} - I_C R_C$

$$V_C = V_{CC} - I_C R_C = 12\ \text{V} - (1.44\ \text{mA})\,(2\ \text{k}\Omega) = 9.12\ \text{V}$$

$$V_{CE} = V_C - V_E = 9.12\ \text{V} - 2.16\ \text{V} = 6.96\ \text{V}$$

AC

- R_{in}: Thinking process: $R_{in} = R_1 \,\|\, R_2 \,\|\, \beta_{ac} r_e \overset{?}{\rightarrow} r_e = \dfrac{25\ \text{mV}}{I_E}$

$$r_e = \frac{25\ \text{mV}}{I_E} = \frac{25\ \text{mV}}{1.44\ \text{mA}} \approx 17.36\ \Omega$$

$$R_{in} = R_1 \parallel R_2 \parallel \beta_{ac} r_e = 15 \text{ k}\Omega \backslash\backslash 4.7 \text{ k}\Omega \backslash\backslash (200)(17.36 \ \Omega) \quad h_{fe} = \beta_{ac}$$

$$\approx 3.579 \text{ k}\Omega \backslash\backslash 3.472 \text{ k}\Omega \approx 1.762 \text{ k}\Omega$$

– V_b: $\quad V_b = V_{in} \dfrac{R_{in}}{R_s + R_{in}} = 8 \text{ mV} \dfrac{1.762 \text{ k}\Omega}{0.1 \text{ k}\Omega + 1.762 \text{ k}\Omega} \approx 7.57 \text{ mV}$

$\quad\quad R_s = 100 \ \Omega = 0.1 \text{ k}\Omega$

– R_o: $\quad R_o \approx R_C = 2 \text{ k}\Omega$

– A_V **without C_E**: $\quad A_V = \dfrac{R_C}{r_e + R_E} = \dfrac{2 \text{ k}\Omega}{17.36 \ \Omega + 1.5 \text{ k}\Omega}$

$$= \dfrac{2 \text{ k}\Omega}{0.01736 \text{ k}\Omega + 1.5 \text{ k}\Omega} \approx 1.32$$

– A_V **with C_E** (without R_L): $\quad A_V = \dfrac{R_C}{r_e} = \dfrac{2 \text{ k}\Omega}{17.36 \ \Omega} = \dfrac{2 \text{ k}\Omega}{0.01736 \text{ k}\Omega} \approx 115.2$

1.32 vs. 115.2 – what a difference a bypass capacitor can make!

– A_V **with C_E and R_L**: $\quad A_V = \dfrac{R_C \backslash\backslash R_L}{r_e} = \dfrac{2 \text{ k}\Omega \ \backslash\backslash 47 \text{ k}\Omega}{17.36 \ \Omega}$

$$= \dfrac{1.92 \text{ k}\Omega}{0.01736 \text{ k}\Omega} = 110.6$$

Phase relationship

- 180^0 phase shift: the output voltage in a C_E amplifier is 180^0 out of phase with the input voltage.

  A waveform that enters the input of the common emitter amplifier will have a 180° phase change at the output.

- When the input voltage V_{in} increases, base current I_b increases. In turn, this increases in collector current I_c. This causes a voltage drop in the collector terminal V_c and produces an 180^0 phase shift.

$$V_c = V_{CC} - I_c R_C \qquad\qquad I_b = \beta_{ac} I_c$$

Current gain

- Amplifier current gain: the ratio of output to input in the form of current.
- The current gain of a BJT (device's current gain) is the ratio of collector current I_c to the base current I_b.

$$\beta_{ac} = h_{fe} = \dfrac{I_c}{I_b}$$

- The current gain for a C_E amplifier is the ratio of collector current I_c to the input current I_{in}.

$$A_i \approx \frac{I_C}{I_{in}}$$

Where $I_{in} \approx \frac{V_{in}}{R_{in} + R_s}$

$R_{in} = R_1 \parallel R_2 \parallel \beta_{ac} r_e$

R_s – source resistance

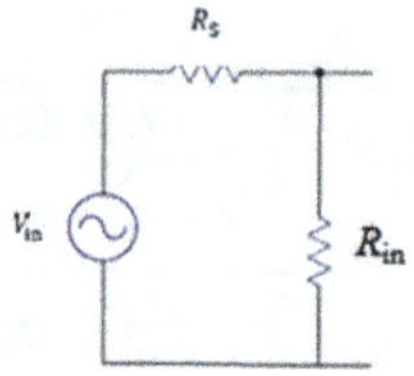

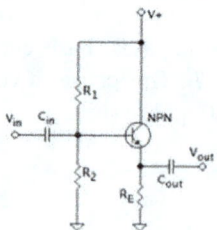

Figure 6.29　Derive the current gain

Power gain

- Amplifier power gain: the ratio of output to input in the form of power.
- Power gain is the product of current gain and voltage gain.

$$A_P = A_v A_i$$

$P = VI$

6.2.2　*Transistor common-collector amplifier*

Common-collector (CC) circuit

- Collector – common
- V_{in} – base
- V_{out} – emitter

Figure 6.30　Common-collector amplifier

AC equivalent circuit

- AC equivalent circuit of a CC amplifier is obtained by replacing all capacitors with a short circuit and setting DC voltage source to zero and replacing it by ground.

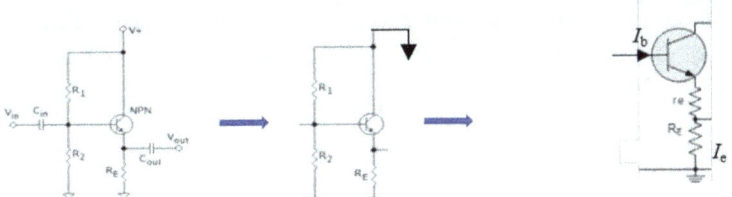

Figure 6.31(a)   CC amplifier

Figure 6.31(b)   AC equivalent circuit

Voltage gain

- Voltage gain without the load resistor R_L:

$$A_V = \frac{R_E}{R_E + r_e}$$

Derive: $A_V = \dfrac{V_{out}}{V_{in}}$ $V_{out} = V_e, \; V_{in} = V_b$

$$A_V = \frac{V_e}{V_b} = \frac{I_e R_E}{I_e (R_E + r_e)} = \frac{R_E}{R_E + r_e}$$

- Voltage gain with the load resistor R_L:

$$A_V = \frac{R_E \backslash\backslash R_L}{R_E \backslash\backslash R_L + r_e}$$

- Voltage gain is approximately 1: usually $R_E \backslash\backslash R_L$ is much larger than r_e and voltage gain A_V is approximately equal to one.

$$A_V \approx 1$$

Derive: $A_V = \dfrac{R_E \backslash\backslash R_L}{R_E \backslash\backslash R_L + r_e} \approx \dfrac{R_E \backslash\backslash R_L}{R_E \backslash\backslash R_L} \approx 1$ If $R_E \backslash\backslash R_L \gg r_e$

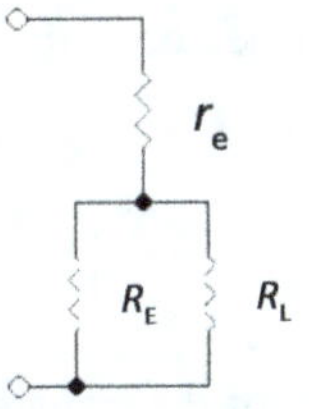

Figure 6.32   Voltage gain with R_L

Phase relationship

- Voltage follower or emitter follower: the CC amplifier is also known as the voltage follower or emitter follower because the load voltage "follows" the input signal very closely and voltage gain is approximately equal to one ($A_V \approx 1$).

- In phase: the CC amplifier receives its input signal to the base with the output voltage taken from across the emitter load. As the emitter voltage follows the base voltage, the output voltage in a CC amplifier is in phase with the input voltage (it simply follows it).

  If a rise in the input voltage causes a rise in the output voltage, V_{out} is in phase with V_{in}, or phase shift is zero.

Current gain

- The current gain for a CC amplifier is the ratio of the emitter current I_e (output) to the input current I_{in} or the base current I_b.

$$A_i = \frac{I_o}{I_{in}} = \frac{I_e}{I_b}$$
$$I_c = \beta_{ac} I_b$$

  - If $R_1 \| R_2 >> \beta_{ac}(R_E \| R_L)$:
    $$A_i \approx \beta_{ac} + 1$$

    Derive: $A_i = \dfrac{I_o}{I_{in}} = \dfrac{I_e}{I_b} = \dfrac{I_c + I_b}{I_b}$ $I_o = I_e,\ I_e = I_c + I_b$

    $$A_i = \frac{I_c}{I_b} + \frac{I_b}{I_b} = \frac{I_c}{I_b} + 1$$

    $$A_i = \frac{\beta_{ac} I_b}{I_b} + 1 = \beta_{ac} + 1$$ $I_c = \beta_{ac} I_b$

  - If $\beta_{ac} >> 1$ $A_i \approx \beta_{ac}$ $A_i = \beta_{ac} + 1$

- The current gain of a common-collector amplifier is high (close to the transistor's current gain β_{ac}).
- The common-collector amplifier is a current amplifier that does not amplify voltage signals ($A_V \approx 1$).

Power gain

- Power gain of the common-collector amplifier is always close to the current gain (since $A_V \approx 1$).

$$A_p = A_V A_i \approx A_i$$ $A_V \approx 1$

Input resistance

- The input resistance at the base of the transistor:
  - R_{ib} without the load resistor R_L:
    $$R_{ib} = \beta_{ac}(r_e + R_E)$$

    Derive: $R_{ib} = \dfrac{V_b}{I_b} = \dfrac{I_e(r_e + R_E)}{I_b} = \dfrac{\beta_{ac} I_b (r_e + R_E)}{I_b} = \beta_{ac}(r_e + R_E)$

    If $R_E >> r_e$: $R_{ib} \approx \beta_{ac} R_E$

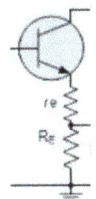

Figure 6.33   input resistance at the base

– R_{ib} with the load resistor R_L:

$$R_{ib} = \beta_{ac}\left(r_e + R_E \backslash\backslash R_L\right)$$

     If $R_E \backslash\backslash R_L \gg r_e$: $R_{ib} \approx \beta_{ac}\left(R_E \backslash\backslash R_L\right)$

- The total circuit input resistance presented to the AC source:

$$R_{in} = R_1 \backslash\backslash R_2 \backslash\backslash R_{ib}$$

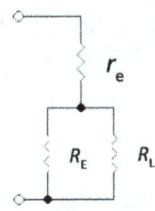

Figure 6.34   Input resistance with R_L

- High input resistance: the value of input resistance on a CC amplifier is usually very high.
  Since R_{ib} is calculated with the current gain β_{ac} of the transistor, and β_{ac} is usually very high. $(R_{ib} \approx \beta_{ac} R_E)$

Output resistance

- The output resistance acts as the source resistance for its load.
- The output resistance R_o presented to the load of the circuit:

$$R_o \approx \left(\frac{R_s}{\beta_{ac}} + r_e\right) \backslash\backslash R_E$$

                                                                      $\beta_{ac} = h_{fe}$

Derive: $R_o = \dfrac{V_o}{I_o} = \dfrac{V_e}{I_e} = \dfrac{V_e}{\beta_{ac} I_b}$

$$I_b = \frac{V_e}{(R_s \backslash\backslash R_1 \backslash\backslash R_2) + r_e}$$

     If R_1 and $R_2 \gg R_s$ and r_e: $I_b \approx \dfrac{V_e}{R_s}$

$$R_o = \frac{V_e}{I_e} = \frac{V_e}{\beta_{ac}I_b} = \frac{V_e}{\beta_{ac}\dfrac{V_e}{R_s}} \qquad\qquad I_e \approx I_c = \beta_{ac}I_b, \ \ I_b = \frac{V_e}{R_s}$$

∵ the load sees R_E in parallel with $\dfrac{R_s}{\beta_{ac}}$, ∴ $R_o \approx \dfrac{R_s}{\beta_{ac}} \backslash\backslash R_E$

If don't neglect r_e (in series with $\dfrac{R_s}{\beta_{ac}}$): $\qquad R_o \approx \left(\dfrac{R_s}{\beta_{ac}} + r_e\right) \backslash\backslash R_E$

- **Low output resistance**: the output resistance on a CC amplifier is always very low.

  Since a CC amplifier can provide high output current gain and without voltage gain, it must have a low output resistance $\left(R \downarrow = \dfrac{V}{I \uparrow}\right)$.

Example: For the BJT CC amplifier of Figure 6.30, determine the input resistance R_{in}, output resistance R_o, and voltage gain A_V, if R_1 is 5 kΩ, R_2 is 4.7 kΩ, R_E is 3.5 kΩ, R_s is 500 Ω, R_L is 20 kΩ, V_{CC} is 12V, $\beta_{DC}= 150$, and h_{fe} is 100. (Assuming a silicon BJT.)

- Given: $R_1 = 5$ kΩ, $R_2 = 4.7$ kΩ, $R_E = 3.5$ kΩ, $R_s = 500$ Ω, $R_L = 20$ kΩ, $V_{CC} = 12$V,

  $\beta_{DC} = 150$, and $h_{fe} = 100$.

- Find: R_{in}, R_o , and A_V.

- Solution:

  $\boldsymbol{R_{in}}$: Thinking process: $R_{in} = R_1 \backslash\backslash R_2 \backslash\backslash \overset{?}{R_{ib}} \ \rightarrow \ R_{ib} \approx \beta_{ac}\overset{?}{\left(R_E\backslash\backslash R_L\right)}$

  $$R_E\backslash\backslash R_L = (3.5 \text{ kΩ}) \backslash\backslash (20 \text{ kΩ}) \approx 2.98 \text{ kΩ}$$

  $$R_{ib} \approx \beta_{ac}\left(R_E\backslash\backslash R_L\right) = (100)(2.98 \text{ kΩ}) = 298 \text{ kΩ} \qquad\qquad \beta_{ac} = h_{fe}$$

  $$\boldsymbol{R_{in}} = R_1 \backslash\backslash R_2 \backslash\backslash R_{ib} = (5 \text{ kΩ}) \backslash\backslash (4.7 \text{ kΩ}) \backslash\backslash (298 \text{ kΩ}) \approx \textbf{2.4 kΩ}$$

  $\boldsymbol{R_o}$: Thinking process: $R_o \approx (\dfrac{R_s}{\beta_{ac}} + \overset{?}{r_e}) \backslash\backslash R_E \rightarrow r_e = \dfrac{25 \text{ mV}}{\overset{?}{I_E} \text{ (mA)}} \rightarrow I_E = \dfrac{\overset{?}{V_E}}{R_E} \rightarrow$

  $$V_E = \overset{?}{V_B} - 0.7\text{V} \ \rightarrow \ V_B \approx V_{CC}\frac{R_2}{R_1 + R_2}$$

  $$V_B \approx V_{CC}\frac{R_2}{R_1 + R_2} = 12 \text{ V}\frac{4.7 \text{ kΩ}}{5 \text{ kΩ} + 4.7 \text{ kΩ}} \approx 5.8 \text{ V} \qquad \beta_{DC}\,R_E \gg R_2$$

  $$V_E = V_B - 0.7 \text{ V} = 5.8 \text{ V} - 0.7 \text{ V} = 5.1 \text{ V}$$

$$I_E = \frac{V_E}{R_E} = \frac{5.1 \text{ V}}{3.5 \text{ k}\Omega} \approx 1.457 \text{ mA}$$ Milli: 10^{-3}; Kilo: 10^{3}

$$r_e = \frac{25 \text{ mV}}{I_E \text{ (mA)}} = \frac{25 \text{ mV}}{1.457 \text{ mA}} \approx 17.16 \text{ }\Omega \approx 0.017 \text{ k}\Omega$$

$$\boldsymbol{R_o} \approx (\frac{R_s}{\beta_{ac}} + r_e) \setminus\setminus R_E = (\frac{500 \text{ }\Omega}{100} + 17.16 \text{ }\Omega) \setminus\setminus 3.5 \text{ k}\Omega$$

$$= 22.16 \text{ }\Omega \setminus\setminus 3500 \text{ }\Omega \approx \textbf{\color{blue}{22.02 }}\boldsymbol{\color{blue}{\Omega}}$$

A_v: $$A_v = \frac{R_E \setminus\setminus R_L}{R_E \setminus\setminus R_L + r_e} = \frac{3.5 \text{ k}\Omega \setminus\setminus 20 \text{ k}\Omega}{3.5 \text{ k}\Omega \setminus\setminus 20 \text{ k}\Omega + 0.017 \text{ k}\Omega}$$

$$\approx \frac{2.98 \text{ k}\Omega}{2.98 \text{ k}\Omega + 0.017 \text{ k}\Omega} \approx \textbf{\color{blue}{0.994}}$$ $A_V \approx 1$

6.2.3   Transistor common-base amplifier

Common-base (CB) circuit

- Base – common
- V_{in} – emitter
- V_{out} – collector

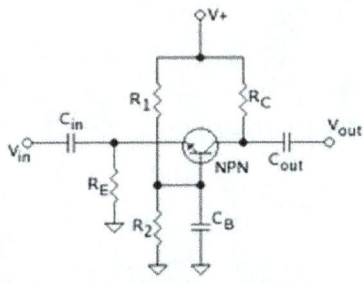

Figure 6.35   Common-base amplifier

AC equivalent circuit

- The AC equivalent circuit for the CB amplifier is obtained by replacing all capacitors with a short-circuit and setting DC voltage sources to zero and replacing them by ground.

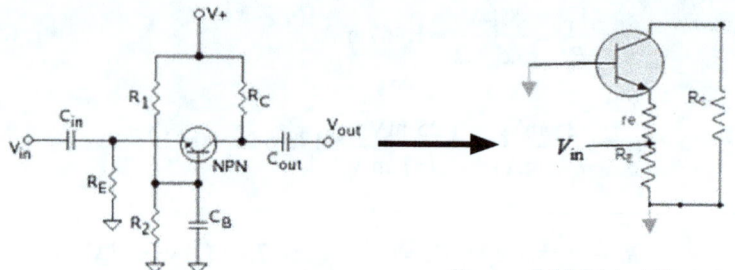

Figure 6.36(b) AC equivalent
circuit of a CB amp

Figure 6.36(a) CB amplifier

Input resistance

- The input resistance at the emitter of the transistor:

$$R_{in} = r_e \backslash\backslash R_E$$

Derive: $R_{in} = \dfrac{V_{in}}{I_{in}} = \dfrac{V_e}{I_e} = \dfrac{I_e (r_e \backslash\backslash R_E)}{I_e} = r_e \backslash\backslash R_E$

- If $R_E \gg r_e$: $R_{in} \approx r_e$

Output resistance

- The output resistance presented to the load (at the collector) of the circuit:

$$R_o = R_C \backslash\backslash r_c$$

Recall: AC collector resistance r_c represents reverse-biased collector-base junction resistance.
- AC collector resistance r_c is typically much larger than R_C and can be replaced by an open circuit.

$$\therefore \quad R_o \approx R_C$$

Voltage gain

- Voltage gain without the load resistor R_L:

$$A_V = \dfrac{R_C}{R_E \backslash\backslash r_e}$$

Derive: $A_V = \dfrac{V_{out}}{V_{in}} = \dfrac{V_c}{V_e} = \dfrac{I_c R_C}{I_e (R_E \backslash\backslash r_e)} \approx \dfrac{I_e R_C}{I_e (R_E \backslash\backslash r_e)} = \dfrac{R_C}{R_E \backslash\backslash r_e}$ $I_c \approx I_e$

If $R_E \gg r_e$: $A_V \approx \dfrac{R_C}{r_e}$

- Voltage gain with the load resistor R_L:

$$A_V = \dfrac{R_C \backslash\backslash R_L}{R_E \backslash\backslash r_e}$$

If $R_E \gg r_e$: $A_V \approx \dfrac{R_C \setminus\setminus R_L}{r_e}$

Current gain

* The current gain for a CB amplifier is the ratio of collector current I_c (output) to the input emitter current I_e.

$$A_i = \frac{I_o}{I_{in}} = \frac{I_c}{I_e} \approx 1$$

$I_c \approx I_e$

Power gain

* The power gain of the CB amplifier is always close to the voltage gain (since $A_i \approx 1$).

$$A_p = A_v A_i \approx A_v$$

$A_i \approx 1$

Phase relationship

* In phase: the output voltage in a CB amplifier is in phase with the input voltage.
* The CB amplifier receives its input signal to the emitter with the output voltage taken from across the collector load ($I_c \approx I_e$, the collector signal "follows" the emitter signal).

6.2.4 *Multistage transistor amplifiers*

Multistage amplifiers

* Multistage amplifier: connect two or more single amplifiers together.
    - Several stages are connected to form a multistage amplifier.
    - The purpose of using a multistage amplifier is to increase the overall gain.
* Cascaded amplifier:
    - A multistage amplifier with the output of one feeds the input of the next.
    - The amplifiers are connected in a left-to-right (in series) horizontal chain configuration.

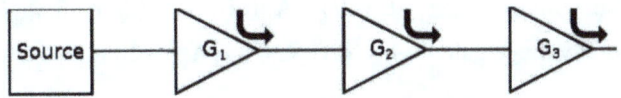

Figure 6.37 Cascaded amplifier

    - A cascaded amplifier uses two or more single-stage common-emitter amplifiers.
* Cascode amplifier:
    - A multistage amplifier that the amplifier is placed one above the other (stacked vertically or in parallel.)

– A cascode amplifier with common emitter as the first stage and common base as second stage.

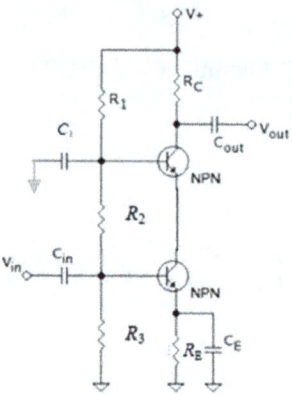

Figure 6.38   Cascode amplifier

Overall gain for a multistage amplifier

- Overall gain: the overall gain for the multistage amplifier is the product of gain of the individual stages.
- Calculation of the overall total gain of the circuit (A_T): the total gain *of* the multistage amplifier *is the* product of the individual gains.

   $$A_T = A_1 A_2 \ldots A_n$$

   A_1 = the gain of the 1st stage
   A_2 = the gain of the 2nd stage
   A_n = the gain of nth stage

- The total gain in logarithmic decibel (dB) units: the amplifier gain is often expressed in logarithmic decibel (dB) units, the total gain is the sum of the gains of the individual gains.

   – The total gain: $A_T\,(\text{dB}) = A_1\,(\text{dB}) + A_2\,(\text{dB}) + \ldots A_n\,(\text{dB})$

   – The total voltage gain: $A_{VT}\,(\text{dB}) = 20 \log A_{V1} + 20 \log_{AV2} + \ldots 20 \log_{AVn}$

   $$\log_a (MN) = \log_a M + \log_a N$$

   or $A_{VT}\,(\text{dB}) = 20 \log_{AVT}$

Coupling capacitor in multistage amplifier

- Blocking DC: a coupling capacitor is used to connect two amplifiers to isolate preceding and succeeding stages.
- Passing AC: a coupling capacitor is used such that only the AC signal from the output of one stage can pass through to the input of its next stage.

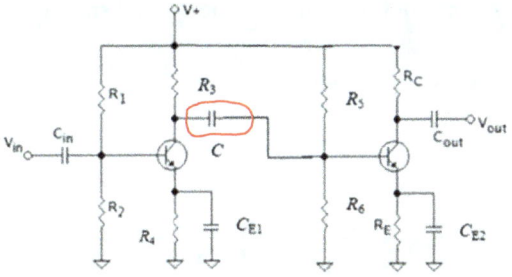

Figure 6.39   Coupling capacitor

Example: For the BJT two-stage amplifier (CE) of Figure 6.39, determine the total overall voltage gain A_{VT} if R_1 is 20 kΩ, R_2 is 4.7 kΩ, R_3 is 3.5 kΩ, R_4 is 1 kΩ, R_5 is 18 kΩ, R_6 is 3 kΩ, R_7 is 4 kΩ, R_8 is 1 kΩ, R_L is 10 kΩ, V_{CC} is 12 V, β_{DC} is 150, and h_{fe} is 200 (for both BJTs). (Assuming two silicon BJTs.)

- Given: $R_1 = 20$ kΩ, $R_2 = 4.7$ kΩ, $R_3 = 3.5$ kΩ, $R_4 = 1$ kΩ, $R_5 = 18$ kΩ, $R_6 = 3$ kΩ, $R_7 = 4$ kΩ, $R_8 = 1$ kΩ, $R_L = 10$ kΩ, $V_{CC} = 12$V, $\beta_{DC} = 150$, and $h_{fe} = 200$.

- Find: A_{VT}

- Solution:

A_{VT}:   Thinking process: $A_{VT} = A_{V1} A_{V2} \rightarrow A_{V1} = \dfrac{R_{C1}^{?}}{r_{e1}^{?}},\quad A_{V2} = \dfrac{R_{C2}^{?}}{r_{c2}^{?}}$

(For a CE amplifier: $A_{V} = \dfrac{R_{C}}{r_{e}}$)

− r_{e1} (the AC emitter resistance of the first stage):

Thinking process: $r_{e1} = \dfrac{25 \text{ mV}}{?I_{E1} \text{ (mA)}} \rightarrow I_{E1} = \dfrac{V_{E1}^{?}}{R_{E1}} \rightarrow V_{E1} = V_{B1}^{?} - 0.7\text{V} \rightarrow$

$V_{B1} \approx V_{CC} \dfrac{R_2}{R_1 + R_2}$

$$V_{B1} \approx V_{CC} \dfrac{R_2}{R_1 + R_2} = 12 \text{ V} \dfrac{4.7 \text{ k}\Omega}{20 \text{ k}\Omega + 4.7 \text{ k}\Omega} \approx 2.28 \text{ V}$$

$$\beta_{DC} R_E \gg R_2,\ (R_E = R_4)$$

$$V_{E1} = V_{B1} - 0.7 \text{ V} = 2.28 \text{ V} - 0.7 \text{ V} = 1.58 \text{ V}$$

$$I_{E1} = \dfrac{V_{E1}}{R_{E1}} = \dfrac{V_{E1}}{R_4} = \dfrac{1.58 \text{ V}}{1 \text{ k}\Omega} = 1.58 \text{ mA}$$

Milli: 10^{-3}; Kilo: 10^{3}

$$r_{e1} = \dfrac{25 \text{ mV}}{I_{E1} \text{ (mA)}} = \dfrac{25 \text{ mV}}{1.58 \text{ mA}} \approx 15.82 \ \Omega = 0.0158 \text{ k}\Omega$$

    — r_{e2} (the AC emitter resistance of the second stage):

Thinking process: $r_{e2} = \dfrac{25\ \text{mV}}{?\,I_{E2}\ (\text{mA})} \rightarrow I_{E2} = \dfrac{V_{E2}^{\ ?}}{R_{E2}} \rightarrow V_{E2} = V_{B2}^{\ ?} - 0.7\ \text{V} \rightarrow$

$$V_{B2} \approx V_{CC}\dfrac{R_6}{R_5 + R_6}$$

$$V_{B2} \approx V_{CC}\dfrac{R_6}{R_5 + R_6} = 12\ \text{V}\dfrac{3\ \text{k}\Omega}{18\ \text{k}\Omega + 3\ \text{k}\Omega} \approx 1.714\ \text{V}$$

$$V_{E2} = V_{B2} - 0.7\ \text{V} = 1.714\ \text{V} - 0.7\ \text{V} \approx 1.014\ \text{V}$$

$$I_{E2} = \dfrac{V_{E2}}{R_{E2}} = \dfrac{V_{E2}}{R_8} = \dfrac{1.014\ \text{V}}{1\ \text{k}\Omega} \approx 1.014\ \text{mA} \qquad\qquad R_{E2} = R_8$$

$$r_{e2} = \dfrac{25\ \text{mV}}{I_{E2}\ (\text{mA})} = \dfrac{25\ \text{mV}}{1.014\ \text{mA}} \approx \mathbf{24.7\ \Omega} = 0.0247\ \text{k}\Omega$$

    — $\boldsymbol{R_{c1}}$ (the AC collector resistance of the first stage):

Thinking process: $R_{c1} = R_3 \parallel R_5 \parallel R_6 \parallel R_{ib2}^{\ ?} \rightarrow R_{ib2} = \beta_{ac}\,r_{e2}$

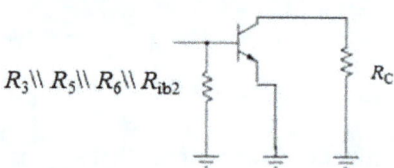

$R_3 \parallel R_5 \parallel R_6 \parallel R_{ib2}$ R_C

Figure 6.40   AC equivalent circuit of a
two-stage amplifier

$$R_{ib2} = \beta_{ac}\,r_{e2} = h_{fe}\,r_{e2} = (200)\,(24.7\ \Omega) = 4940\ \Omega \approx 4.94\ \text{k}\Omega \qquad \beta_{ac} = h_{fe}$$

$$\boldsymbol{R_{c1}} = R_3 \parallel R_5 \parallel R_6 \parallel R_{ib2} = 3.5\ \text{k}\Omega \parallel 18\ \text{k}\Omega \parallel 3\ \text{k}\Omega \parallel 4.94\ \text{k}\Omega$$
$$\approx \mathbf{1.14\ k\Omega}$$

  — $\boldsymbol{A_{V1}}$: $A_{V1} = \dfrac{R_{c1}}{r_{e1}} = \dfrac{1.14\ \text{k}\Omega}{0.0158\ k\Omega} \approx \mathbf{72.15}$

  — $\boldsymbol{A_{V2}}$:   Thinking process: $A_{V2} = \dfrac{R_{c2}^{\ ?}}{r_{e2}} \rightarrow R_{c2} = R_7 \parallel R_L$

$$A_{V2} = \dfrac{R_{c2}}{r_{e2}} = \dfrac{R_7 \parallel R_L}{r_{e2}} = \dfrac{4\ \text{k}\Omega \parallel 10\ \text{k}\Omega}{0.0247\ k\Omega} \approx \dfrac{2.857\ \text{k}\Omega}{0.0247\ k\Omega} \approx \mathbf{115.7}$$

$$- \quad A_{VT}: \quad A_{VT} = A_{V1} A_{V2} = (72.15)(115.7) \approx 8348$$

$$A_{VT} \textbf{(dB)} = A_1 \text{(dB)} + A_2 \text{(dB)} = 20 \log (A_{V1}) + 20 \log (A_{V2})$$

$$= 20 \log (72.15) + 20 \log (115.7) \approx 37.17 \text{ dB} + 41.27 \text{ dB}$$

$$\approx 78.44 \text{ dB}$$

Summary

Amplifier gain:

- Power gain:

$$A_P = \frac{P_{out}}{P_{in}}$$

- Voltage gain:

$$A_V = \frac{V_{out}}{V_{in}}$$

- Current gain:

$$A_i = \frac{i_{out}}{i_{in}}$$

The function of capacitors in BJT amplifier

- Capacitor: pass AC signal (short circuit) and block DC (open circuit).
- Coupling capacitors
  - Coupling capacitor: a capacitor that is connected between two nodes of a circuit such that only the AC signal can pass through while DC is blocked (a DC blocking capacitor).
  - Bypass capacitor: a capacitor that is used to short AC signal to the ground so that the AC noise is removed. It also acts as an open circuit for a DC and maintains the DC bias.

Superposition of DC and AC waveforms

- Superposition: two or more waves combine in a manner which is the algebraic sum of the waveforms produced by each independent wave acting separately.
- Superposition of AC and DC: combine the waveforms of the AC and DC.

Transistor AC model

- *r*-parameter or *h*-parameters model: an AC equivalent circuit that can be used to accurately predict the performance of a BJT. It can be used to quickly estimate the input impedance, gain, and operating conditions of BJT amplifiers.
- Simplified transistor *r* parameter AC equivalent circuit:

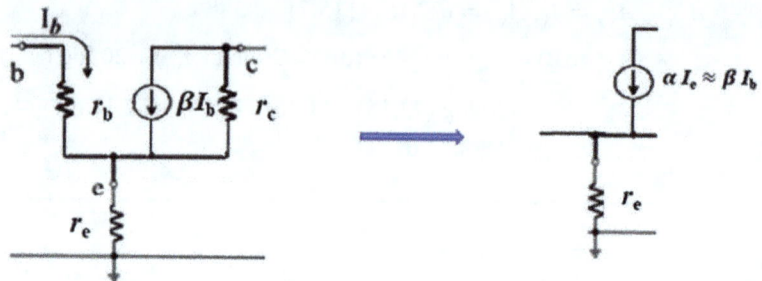

Figure 6.41   Simplified BJT r parameter equivalent circuit

- BJT internal AC resistance r: a small AC resistance looking into the terminal of a transistor.

Table 6.6   r parameters

r parameter	Formula
AC emitter resistance r_e	$r_e = \dfrac{26 \text{ mV}}{I_E \text{ (mA)}}$ (at room temperature)
AC base resistance r_b	very small – short circuit
AC collector resistance r_c	very large – open circuit
AC current gain β_{ac}	$\beta_{ac} = \dfrac{I_c}{I_b}$
AC current gain α_{ac}	$\alpha_{ac} = \dfrac{I_c}{I_e}$

Transistor *h* parameters

- Hybrid parameters or *h* parameters: every linear circuit (or network) having input and output terminals can be analyzed by four parameters called hybrid (*h*) parameters.
- Hybrid parameters h_f and h_r: they are the same as the *r* parameters (AC current gain β_{ac} or α_{ac}) but are more widely used in transistor datasheets.
- Small-signal analysis: both *r* and *h* parameters are valid only for small-signal analysis (the amplifier's linear region of operation).

Table 6.7   h *parameters*

h parameters	Subscript	Condition	Formula	Unit
Input impedance or resistance h_i	i – input	Output short circuit	$h_i = \dfrac{V_i}{I_i}$	ohm
Output admittance or conductance h_o	o – output	Input open circuit	$h_o = \dfrac{I_o}{V_o}$	mho
Forward current gain h_f	f – forward	Output short circuit	$h_f = \dfrac{I_o}{I_i}$	dimensionless
Reverse voltage gain h_r	r – reverse	Input open circuit	$h_r = \dfrac{V_i}{V_o}$	dimensionless

Fundamental BJT amplifier circuit configurations

- There are three most fundamental configurations for a transistor amplifier.

Table 6.8   *Types of amplifier configurations*

Amplifier	Abbreviation	Input	Output	Common terminal	Subscript
Common-emitter amplifier	CE	base	collector	emitter	e
Common-collector amplifier	CC	base	emitter	collector	c
Common-base amplifier	CB	emitter	collector	base	b

Table 6.9   h *parameters for three configurations*

Configuration	Subscript	h_i	h_o	h_f	h_r
Common-emitter (CE)	e	$h_{ie} = \dfrac{V_b}{I_b}$	$h_{oe} = \dfrac{I_c}{V_{ce}}$	$h_{fe} = \dfrac{I_c}{I_b}$	$h_{re} = \dfrac{V_b}{V_c}$
Common-collector (CC)	c	$h_{ic} = \dfrac{V_b}{I_b}$	$h_{oc} = \dfrac{I_e}{V_e}$	$h_{fc} = \dfrac{I_e}{I_b}$	$h_{rc} = \dfrac{V_b}{V_e}$
Common-base (CB)	b	$h_{ib} = \dfrac{V_e}{I_e}$	$h_{ob} = \dfrac{I_c}{V_c}$	$h_{fb} = \dfrac{I_c}{I_e}$	$h_{rb} = \dfrac{V_e}{V_c}$

Conversion between *h* and *r* parameters

$$h_{fe} = \beta_{ac} = \frac{I_c}{I_b}$$

Multistage amplifiers

- Multistage amplifier: connect two or more single amplifiers together.
- Cascaded amplifier:
  - The amplifiers are connected in a left-to-right (in series) horizontal chain configuration.
  - A cascaded amplifier uses two or more single-stage common-emitter amplifiers.
- Cascode amplifier:
  - A multistage amplifier that the amplifier is placed one above the other.
  - A cascode amplifier with common emitter as the first stage and common base as second stage.

Overall gain for a multistage amplifier

- Calculation of the overall total gain of the circuit A_T: $A_T = A_1 A_2 \dots A_n$
- The total gain in dB: the amplifier gain is often expressed in logarithmic *decibel (dB) units*.
  - The total gain: A_T (dB) = A_1 (dB) + A_2 (dB) + ... A_n (dB)
  - The total voltage gain: A_{VT} (dB) = 20 log A_{V1} + 20 log $_{AV2}$ + ... 20 log $_{AVn}$
          or A_{VT} (dB) = 20 log $_{AVT}$

Coupling capacitor in multistage amplifier

- Blocking DC: a coupling capacitor is used to connect two amplifiers to isolate preceding and succeeding stages.
- Passing AC: a coupling capacitor is used such that only the AC signal from the output of one stage can pass through to the input of its next stage.

Table 6.10 The three basic configurations of BJT amplifier

Configuration	A_v	A_i	A_p	R_{in}	R_o	Phase relationship
Common Emitter (CE)	$\dfrac{R_C}{r_e}$	$\dfrac{I_c}{I_{in}}$ $I_{in} \approx \dfrac{V_s}{R_{in}+R_s}$	$A_v A_i$	$R_1 \| R_2 \| \beta_{ac}\, r_e$	$R_C \| r_c \approx R_C$	180^0 out of phase
Common Collector (CC)	1	$\beta_{ac}+1$	$A_v A_i \approx A_i$	$R_1 \| R_2 \| R_{ib}$ $R_{ib}=\beta_{ac}\left(r_e+R_e\right)$	$R_o \approx \left(\dfrac{R_s}{\beta_{ac}}\right) \| R_E$	In phase
Common Base (CB)	$\dfrac{R_C}{R_E \| r_e} \approx \dfrac{R_C}{r_e}$	1	$A_v A_i \approx A_v$	$r_e \| R_E$	$R_C \| r_c \approx R_C$	In phase

Common Emitter (CE)
Input − base
Output − collector

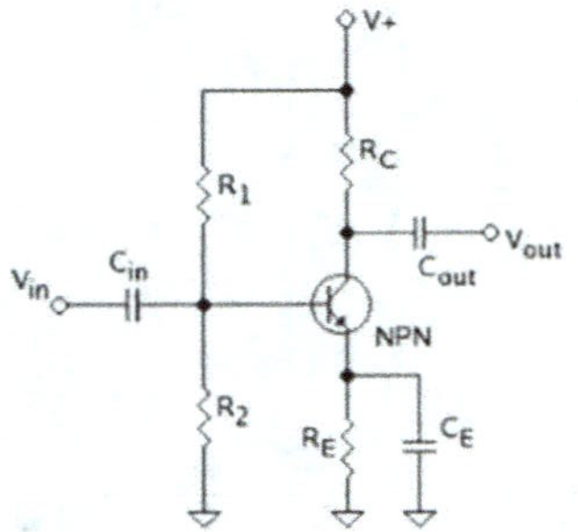

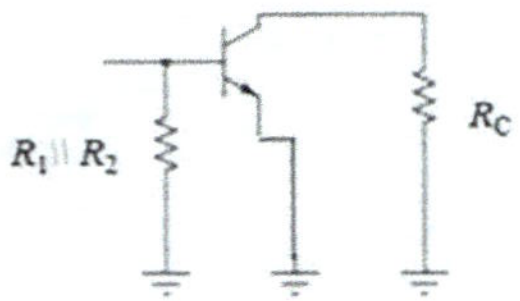

$$I_C \approx I_E = \frac{V_E}{R_E}$$

$$V_{CE} = V_C - V_E$$

$$V_C = V_{CC} - I_C R_C$$

$$V_E = V_B - V_{BE}$$

$$V_B = V_{CC}\,\frac{R_2 \| R_{IN(base)}}{R_1 + R_2 \| R_{IN(base)}}$$

$$\left(R_{IN(base)} \approx \beta_{DC}\, R_E\right)$$

If $R_{IN(base)} \gg R_2$:

$$V_B \approx V_{CC}\,\frac{R_2}{R_1 + R_2}$$

(Continued)

Table 6.10 The three basic configurations of BJT amplifier (Continued)

Configuration	A_v	A_i	A_p	R_{in}	R_o	Phase relationship
Common Collector (CC) Input – base Output – emitter	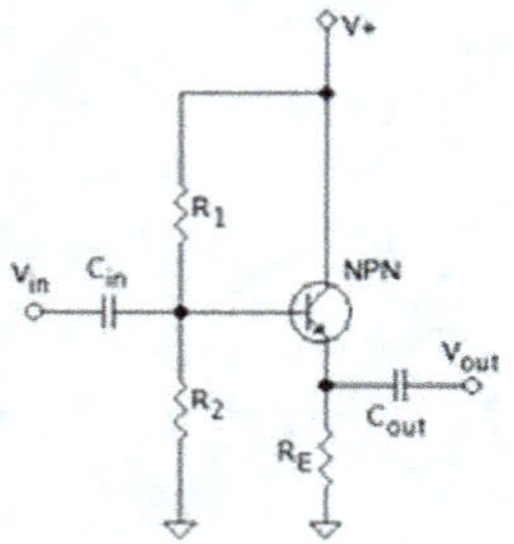		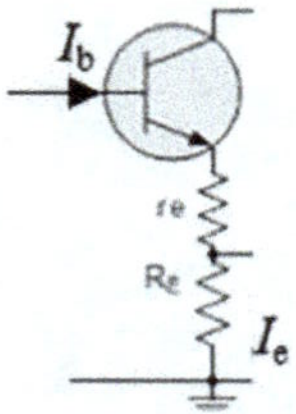		$I_C \approx I_E = \dfrac{V_E}{R_E}$ $V_{CE} = V_C - V_E$ $V_C = V_{CC}$ $V_E = V_B - V_{BE}$ $V_B \approx V_{CC}\dfrac{R_2}{R_1 + R_2}$	
Common Base (CB) Input – emitter Output – collector	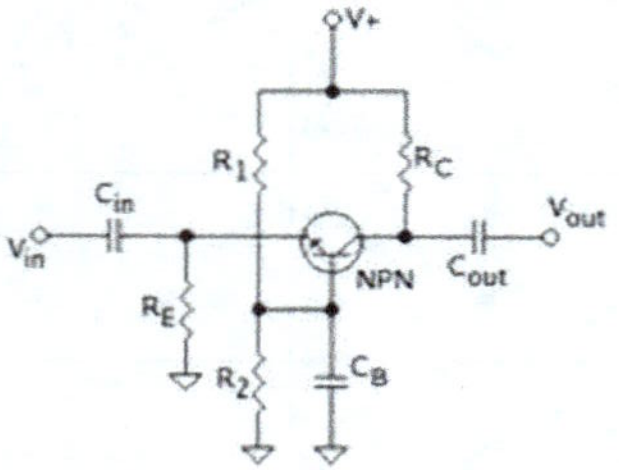		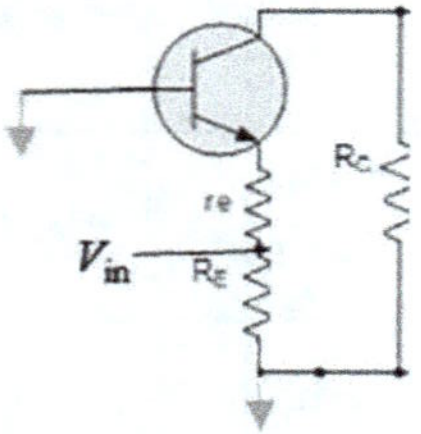		$I_C \approx I_E = \dfrac{V_E}{R_E}$ $V_{CE} = V_C - V_E$ $V_C = V_{CC} - I_C R_C$ $V_E = V_B - V_{BE}$ $V_B \approx V_{CC}\dfrac{R_2}{R_1 + R_2}$	

Self-Test

6.1 1. A transistor AC model can be represented by r parameters or h parameters equivalent circuit. The circuit uses (   ) parameters to represent the BJT's operation.

2. The AC current gain alpha (α_{ac}) is the ratio of the AC collector current to the AC (   ) current.

3. Four hybrid parameters measured in ohm, (   ), and two dimensionless.

4. Hybrid parameters are the same as the r parameters but are more widely used in transistor (   ).

5. Both r and h parameters are valid only for (   )-signal analysis.

6. Common-collector amplifier (CC): the input signal is applied to the (   ) branch of the BJT, the output is taken from the emitter branch of the BJT, and the collector terminal is a common connection point for both input and output.

7. The forward current gain h_{fe} is the same as the (   ) that can be seen from h and r parameters common-emitter equivalent circuits.

8. The input resistance or impedance h_{ib} is the same as the (   ).

6.2 9. The DC equivalent circuit of a CE amplifier is obtained by replacing all capacitors by an (   ) circuit.

10. The AC equivalent circuit of an amplifier is obtained by replacing all capacitors by a (   ) circuit and setting DC voltage sources to zero and replacing them by ground.

11. When a bypass capacitor C_E is connected with an emitter resistance R_E, the voltage gain of CE amplifier (   ).

12. For the BJT amplifier of Figure 6.42, determine the Q-point, input and output resistance R_{in} and R_{out}, AC signal voltage at the base V_b, and voltage gain A_V both with and without the bypass capacitor C_E and load R_L, if R_1 is 10 kΩ, R_2 is 3 kΩ, R_C is 1 kΩ, R_E is 1 kΩ, R_S is 80 Ω, R_L is 40 kΩ, V_{CC} is 10 V, β_{DC} is 75, V_{in} is 6 mV, β_{DC} is 75, and h_{fe} is 150. (Assuming a silicon BJT.)

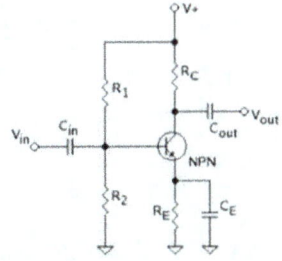

Figure 6.42   Ch 6: No. 12, self-test

13. The output voltage in a CE amplifier is (   ) phase with the input voltage.

14. The common (   ) amplifier is also known as the voltage-follower or emitter-follower.

15. The value of (   ) resistance on a CC amplifier is usually very high.

16. For the BJT CC amplifier of Figure 6.43, determine the input resistance R_{in}, output resistance R_o, and voltage gain A_V, if R_1 is 4 kΩ, R_2 is 3 kΩ, R_E is 2.5 kΩ, R_S is 400 Ω, R_L is 1.5 kΩ, V_{CC} is 10 V, V_{in} is 6 mV, and h_{fe} is 150. (Assuming a silicon BJT.)

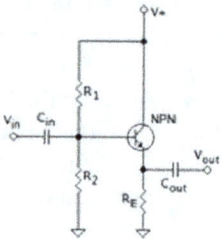

Figure 6.43 Ch 6: No. 16, self-test

17. The purpose of using a multistage amplifier is to increase the overall (   ).

18. The cascaded amplifier uses two or more single-stage common-(   ) amplifiers.

19. The (   ) amplifier is a multistage amplifier that the amplifier is placed one above the other.

20. For the BJT two-stage amplifier (CE) of Figure 6.44, determine the total overall voltage gain A_{VT}, if R_1 is 10 kΩ, R_2 is 3 kΩ, R_3 is 2.5 kΩ, R_4 is 0.5 kΩ, R_5 is 15 kΩ, R_6 is 2 kΩ, R_7 is 3 kΩ, R_8 is 0.8 kΩ, R_L is 8 kΩ, V_{CC} is 15 V, and h_{fe} is 100 for both BJTs. (Assuming a silicon BJT.)

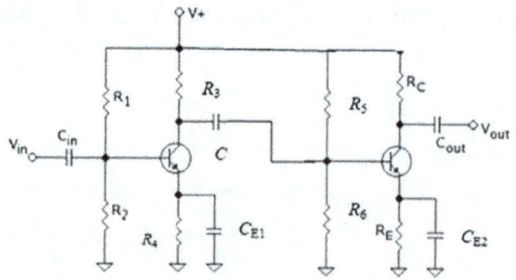

Figure 6.44 Ch 6: No. 20, self-test

Chapter 7

Field-effect transistors

Chapter outline

7.1   Field-effect transistors

7.1.1   Introduction to field-effect transistors

Introduction to FETs

- Field-effect transistor (FET): a type of transistor that uses the value of an electric field to control the output current (a voltage-controlled device). They are used for switching and amplification in circuits.

Table 7.1   The main difference between FET and BJT

FET (Field-effect transistor)	BJT (bipolar junction transistor)
Only the majority charge carriers (holes or electrons) flow	Both majority and minority charge carriers flow
A unipolar device (only one carrier type)	A bipolar device
A voltage-control device	A current-controlled device
High input resistance	Low input resistance
Low power consumption	High power consumption
Low noise level	High noise level

Figure 7.1(a)   Current-controlled device   Figure 7.1(b)   Voltage-controlled device

Types of FET

- Recall – types of BJT: there are two general types of BJTs – PNP and NPN.
- There are two general types of FETs: JFET (Junction Gate Field-Effect Transistor) and MOSFET (Metal–Oxide–Semiconductor Field-Effect Transistor).
    - JFET: a semiconductor device with a PN junction that operates in only depletion mode.
    - MOSFET: a semiconductor device without a PN junction that operates in both depletion and enhancement mode.

Basic structure of JFET

- Types of JFET: there are two major types of JFET – N-channel and P-channel
- N- or P-channel: a JFET consists of a channel of N- or P-type semiconductor material through which current flows.
    - N-channel JFET: a JFET whose channel is composed of a majority of electrons as the charge carrier. It made up a bar of N-type semiconductor with two parts of P-type material inserted in the sides.
    - P-channel JFET: a JFET whose channel is composed of a majority of holes as the charge carrier. It made up a bar of P-type material with two parts of N-type material inserted in the sides.

    Channel: the section of semiconductor material through which the current flows (a conduction path).

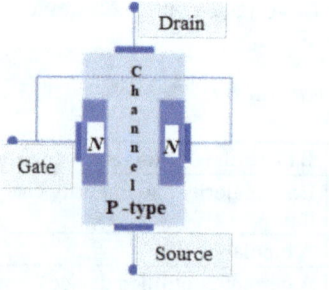

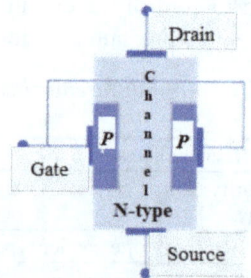

Figure 7.2(a)   P-channel JFET   *Figure 7.2(b)   N-channel JFET*

JFET symbol

N-channel   P-channel

Figure 7.3   FET symbol

- N-channel JFET: the arrow pointing toward the channel (the arrow is pointing in).
- P-channel JFET: the arrow pointing away from the channel (the arrow is pointing out).

Three terminals: the JFET is a three-terminal semiconductor device.
- Drain (D): at the upper end.
- Source (S): at the lower end.
- Gate (G): surrounds the channel (like a belt surrounding the waist) and controls the carrier's flow.

JFET – a voltage-controlled device
- The gate (G) is used to control current flow between the other two terminals - source (S) and drain (D).
- Reverse bias voltage V_{GS}: by applying a reverse bias voltage V_{GS} to the gate-source junction can control the width of the channel and the flow of current.
- Voltage-controlled device: under reverse bias gate voltage V_{GS}, a depletion layer is formed in the channel. As reverse bias changes the width of the depletion layer and the resistance of the channel, so the drain current I_D changes.
  - I_D is controlled by the channel width (the resistance or cross-sectional area of the channel).
  - The channel width is controlled by the depletion layer.
  - The depletion layer is controlled by the reverse bias voltage V_{GS}.

$V_{GS} \rightarrow$ depletion layer $\rightarrow$ channel width $\rightarrow$ resistance of the channel $\rightarrow I_D$

(Cross-sectional area changes)

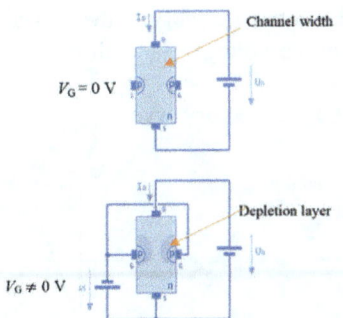

Figure 7.4   Voltage-controlled device

The V_{GS} of a JFET

Since a JFET must be operated such that the gate-source junction is always reverse-biased. This requires a negative V_{GS} for an N-channel JFET and a positive V_{GS} for a P-channel JFET.

- $- V_{GS}$: a N-channel JFET requires a negative V_{GS}.
- $+ V_{GS}$: a P-channel JFET requires a positive V_{GS}.

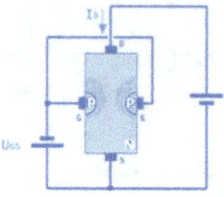

Figure 7.5(a) V_{GS} for a P-channel JFET   Figure 7.5(b) V_{GS} for a N-channel JFET

7.1.2   JFET – operation and characteristics

DC source voltages of an N-channel JFET

- V_{DD}: provides the forward-biased drain-to-source voltage (V_{DS}) and current from drain to source (I_D).
- V_{GG}: provides the reverse-bias voltage between the gate and the source.

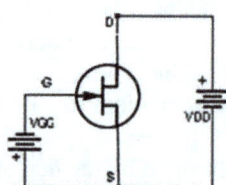

Figure 7.6   DC source voltages of JFET

JFET characteristics curve

- Drain (or output) characteristic curve: the curve plot between the drain current I_D and drain-source voltage V_{DS} with gate-to-source voltage V_{GS} as the parameter.   A plot of V_{DS} verses I_D.
- JFET drain curve vs. BJT collector curve:

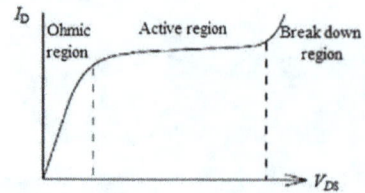

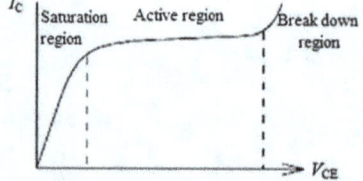

Figure 7.7(a)   JFET drain curve   Figure 7.7(b)   JBJT collector curve

Notice that the similarities between the characteristic curves of the JEFT and BJT.

JFET operation when $V_{GS} = 0$

- Initial operation: when the reverse bias voltage V_{GS} is zero and there is no drain-source voltage V_{DS} applied, the drain current I_D is also zero.

$$V_{GS} = V_{DS} = 0 \rightarrow I_D = 0$$

- Ohmic region (linear region): as the drain-source voltage V_{DS} is increased, the drain current I_D increases (with $V_{GS} = 0$), and JFET acts as a voltage-controlled resistor.

"Ohmic" region: V_{DS} and I_D are related by Ohm's law.

$$V_{DS} \uparrow \rightarrow I_D \uparrow$$

- Active region (or saturation or pinch-off region): the drain current I_D is almost independent of the drain-source voltage V_{DS} (the V_{DS} has little or no effect).
  - The JFET acts as a good conductor and is controlled by the V_{GS}.
  - When the JFET is working as an amplifier, the JFET uses the active region for operation.

I_D approaches a saturation value (JFET is acting like a saturated BJT).

$$V_{DS} \uparrow\uparrow \rightarrow I_D \approx \text{constant} \qquad \qquad V_{GS} = 0$$

- Pinch off voltage V_P: the minimum value V_{DS} at which maximum drain current I_D can flow (any increase in V_{DS} does not cause an increase in I_D – the current is turned to be essentially constant). $V_{DS} \uparrow > V_P \rightarrow I_D \approx \text{constant}$ $V_{GS} = 0$

  o   Operating with V_{DS} below V_P is in the ohmic region. $V_{DS} < V_P$: ohmic region
  o   Operating with V_{DS} above V_P will enter the active region. $V_{DS} > V_P$: active region
  o   Pinch off is where the device crosses from being ohmic to saturating.

- Drain current for zero bias I_{DSS} (or the drain saturation current at $V_{GS} = 0$): the maximum saturation (or steady state) drain current I_D at zero V_{GS}.

I_{DSS} subscript "D" drain current; subscript "SS" steady state.

$$V_{DS} \uparrow\uparrow > V_P \rightarrow I_D = I_{DSS} \qquad \qquad V_{GS} = 0$$

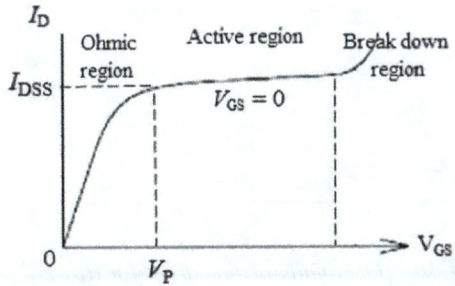

Figure 7.8   JFET characteristics curve

- Breakdown region: the voltage V_{DS} increases too much for the channel to handle, the JFET loses its ability to resist the current which causes the channel to breakdown and the I_D drastically increases.　　　　$V_{DS}\uparrow\uparrow\uparrow \rightarrow I_D\uparrow\uparrow\uparrow$

JFET operation when $V_{GS}\neq 0$

- Voltage controlled JFET: if apply a small voltage V_{GS} to the gate, the width of the depletion layer will increase and narrow the channel. The resistance of the channel will increase and hence the drain current I_D will reduce.

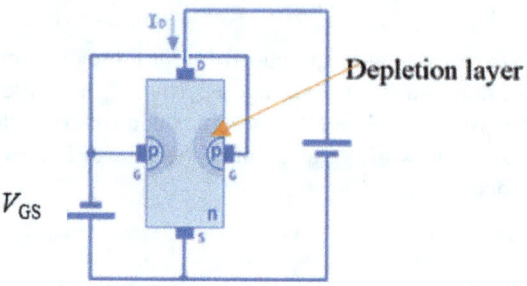

Figure 7.9　Voltage-controlled JFET

Apply a $V_{GS}\rightarrow$ depletion layer width $\uparrow\rightarrow$ channel width $\downarrow\rightarrow$ channel resistance $\uparrow$ $\rightarrow I_D\downarrow$

- The family drain curves for a JFET: if V_{GS} is set to different values, the drain current I_D will have different values. The relationship between V_{GS} and I_D can build a family of characteristic curves for the JFET.

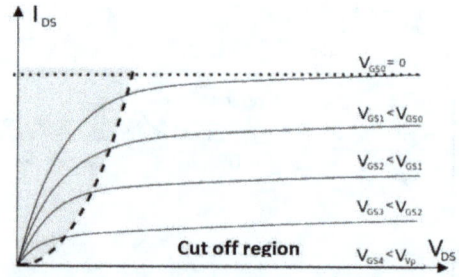

Figure 7.10　The family drain curves for a JFET

For an N-channel JFET: the more negative value V_{GS} is, the smaller I_D in the active region.

$$|V_{GS}|\uparrow \rightarrow I_D\downarrow$$

- Cut-off region (or pinch-off region): the reverse bias voltage V_{GS} is high enough to increase the width of the depletion layer, which causes the channel to close and blocks the flow of the current I_D through the channel.

$$V_{GS} \uparrow\uparrow \rightarrow \text{depletion layer} \uparrow\uparrow \rightarrow \text{channel closed} \rightarrow I_D \approx 0$$

  - Cut-off voltage $V_{GS\,(off)}$: the minimum value of V_{GS} to turn the JFET off. At the voltage $V_{GS\,(off)}$, the JFET enters the cut-off region.
  - At cut-off, JFET acts as an open switch (the channel resistance is at maximum).

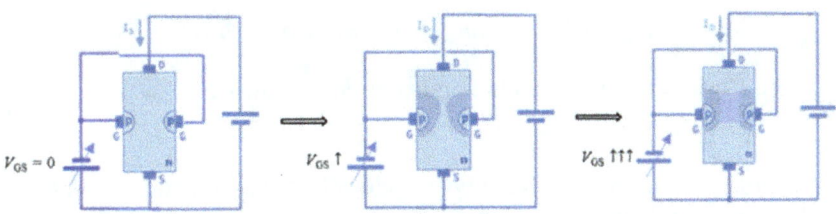

Figure 7.11   JFET cut-off

Pinch-off voltage vs. cut-off voltage

- Pinch-off voltage V_P: the minimum value of V_{DS} at which I_D becomes saturation ($I_D = I_{DSS}$) at zero V_{GS}.

$$V_P: \text{the minimum } V_{DS} \quad \rightarrow \quad I_D = I_{DSS} \hspace{3cm} V_{GS} = 0$$

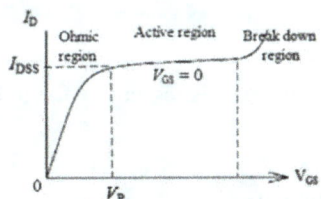

Figure 7.12   Pinch-off voltage V_P

- Cut-off voltage $V_{GS(off)}$: the minimum value of V_{GS} at which I_D becomes approximately zero.

$$V_{GS(off)}: \text{the minimum } V_{GS} \quad \rightarrow \quad I_D \approx 0$$

- V_{GS} and V_P are equal in manganite and opposite in direction.

$$V_{GS} = -V_P$$

Transfer characteristic of JFET (N-channel)

- Transfer characteristic (or transconductance curve): the curve plot between gate-source voltage V_{GS} and drain current I_D (by keeping drain-source voltage V_{DS} at pinch-off voltage V_P). A plot of V_{GS} versus I_D.

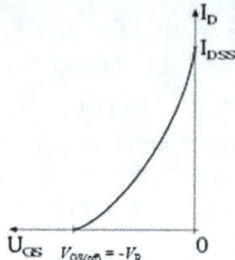

Figure 7.13   Transfer characteristic of JFET

- When V_{GS} is zero, the maximum drain current I_D flowing is I_{DSS}.

    When $V_{GS} = 0$: $I_D = I_{DSS}$

- When $V_{GS} = -V_P$, the transfer curve develops from $I_D = I_{DSS}$ at $V_{GS} = 0$, to $I_D = 0$ at $V_{GS} = -V_P$ (N-channel).

    When $V_{GS} = -V_P$: $I_D = 0$

- Shockley's equation: the transfer characteristic of JFET is defined by Shockley's equation (to calculate drain current I_D or plot a transfer curve).

$$I_D = I_{DSS} \left(1 - \frac{V_{GS}}{V_{GS(off)}}\right)^2$$

The squared term produces a nonlinear exponential relationship between V_{GS} and I_D.

Example: Plot the transfer characteristics curve for a JFET, if I_{DSS} is 3 mA and $V_{GS(off)}$ is negative 6 V.

- Given: $I_{DSS} = 3$ mA, $V_{GS(off)} = -6$ V
- Plot: the transfer curve

- Solution: $I_D = I_{DSS} \left(1 - \frac{V_{GS}}{V_{GS(off)}}\right)^2$

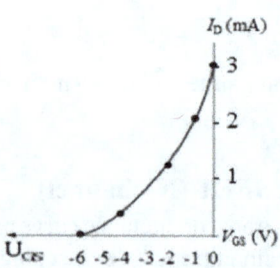

Figure 7.14   Transfer characteristic – an example

Table 7.2   An example

V_{GS}	$I_D = I_{DSS}\left(1 - \dfrac{V_{GS}}{V_{GS(\text{off})}}\right)^2$	Ordered Pair (V_{GS}, I_D)
0	$I_D = (3 \text{ mA})\left(1 - \dfrac{0}{-6 \text{ V}}\right)^2 = 3 \text{ mA}$	(0, 3)
−1 V	$I_D = (3 \text{ mA})\left(1 - \dfrac{-1 \text{ V}}{-6 \text{ V}}\right)^2 \approx 2.08 \text{ mA}$	(−1, 2.08)
−2 V	$I_D = (3 \text{ mA})\left(1 - \dfrac{-2 \text{ V}}{-6 \text{ V}}\right)^2 \approx 1.33 \text{ mA}$	(−2, 1.33)
−4 V	$I_D = (3 \text{ mA})\left(1 - \dfrac{-4 \text{ V}}{-6 \text{ V}}\right)^2 \approx 0.33 \text{ mA}$	(−4, 0.33)
−6 V	$I_D = (3 \text{ mA})\left(1 - \dfrac{-6 \text{ V}}{-6 \text{ V}}\right)^2 = 0 \text{ mA}$	(−6, 0)

Connect the calculated points from the table to get the transfer curve as shown in Figure 7.14.

7.2   Biasing of JFET

7.2.1   *JFET gate bias (or fixed bias)*

The purpose of JFET biasing

- Recall–biasing: establishing a predetermined level of DC voltages or currents in an amplifier so that the AC signal is properly amplified.
- JFET must be properly biased to establish a known Q-point to operate correctly and produce the desired amplification effect.   It is similar to BJT biasing circuits.
  - Q-point of JFET: V_{GS} and I_D
  - Recall: Q-point of BJT: I_C and V_{CE}
- Correct DC biasing of the JFET also establishes its initial AC operating region with an undistorted output signal (amplified signal).

The commonly used methods of JFET biasing

- Gate bias
- Voltage-divider bias
- Self-bias

Gate bias (or fixed bias)

- Gate-bias circuit (N-channel JFET):

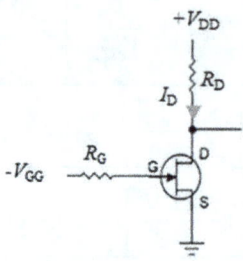

Figure 7.15   Gate-bias circuit

- The gate-bias is obtained using two power supplies $(-V_{GG}$ and $V_{DD})$ – a disadvantage.

  This method is also very unstable to temperature stability.

DC voltages and currents

- Gain-source voltage V_{GS}: the negative power supply $(-V_{GG})$ ensures that the gate is always reverse biased and no current flows through R_G $(I_G = 0)$ so V_{GS} remains constant at $-V_{GG}$.

$$V_{GS} = -V_{GG} \qquad\qquad I_G = 0, \;\; I_G R_G = 0$$

- Drain current I_D: I_D is determined by the transfer characteristic or the Shockley's equation of JFET.

$$I_D = I_{DSS} \left(1 - \frac{V_{GS}}{V_{GS(off)}}\right)^2$$

- Drain-source voltages V_{DS}: $\qquad V_{DS} = V_{DD} - I_D R_D$ $\qquad\qquad$ KVL

7.2.2   JFET Self-bias

Self-bias circuit (N-channel JFET):

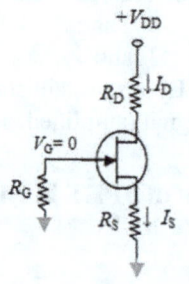

Figure 7.16   Self-bias circuit

- The self-bias is a commonly used method for biasing a JFET.
- There is no voltage across the gate resistor R_G.

$$\because V_G = 0 \qquad \therefore V_{RG} = 0$$

- The gate-source junction remains reverse-biased in the self-bias JFET circuit.

V_G is more negative than V_s or $V_G < V_s$.

Calculating DC voltages and currents

- Drain current I_D: $\qquad I_D \approx I_S$
- Source voltage V_S: $\qquad V_S = I_D R_S$
- Gate-source voltage V_{GS}: $\qquad V_{GS} = -I_D R_S \qquad\qquad I_D I_S = -\dfrac{V_{GS}}{R_S}$

  Derive: $V_{GS} = V_G - V_S = 0 - I_D R_S = -I_D R_S$

- Drain-source voltages V_{DS}: $\qquad V_{DS} = V_{DD} - I_D(R_D + R_S)$

  KVL: $V_{DD} = I_D R_D + V_{DS} + I_S R_S$

Example: Determine the values of V_{GS} and V_{DS} for the circuit shown in Figure 7.16, if R_D is 2 kΩ, R_S is 500 Ω, V_{DD} is 12 V, and I_D is 3 mA.

- Given: $\qquad R_D = 2$ kΩ, $R_S = 500$ Ω, $R_G = 5$ MΩ, $V_{DD} = 12$ V, and $I_D = 3$ mA.
- Find: $\qquad V_{GS}$ and V_{DS}.
- Solution:

  V_{GS}: $\quad V_{GS} = -I_D R_S = -(3 \text{ mA})(500 \text{ Ω}) = -(3 \text{ mA})(0.5 \text{ kΩ}) = -1.5 \text{ V}$

  Milli: 10^{-3}; Kilo: 10^3

  V_{DS}: $\quad V_{DS} = V_{DD} - I_D(R_D + R_S) = 12 \text{ V} - 3 \text{ mA}(2 \text{ kΩ} + 0.5 \text{ kΩ}) = 4.5 \text{ V}$

DC load line and Q-point in self-biasing

- Self-bias JFET DC load line: a line segment drawn on the transfer characteristics curve by connecting the two points (two distinct points determine exactly one line segment).
  - The first point: when $I_D = 0$, $V_{GS} = 0$

    $(V_{GS}, I_D) = (0, 0)$ $\qquad$ The origin of graph

    Derive: if $I_D = 0$, $V_{GS} = -I_D R_S = (0)(R_S) = 0$

  - The second point: when $I_D = I_{DSS}$, $V_{GS} = -I_{DSS} R_S$ $\qquad V_{GS} = -I_D R_S$

    $(V_{GS}, I_D) = (-I_{DSS} R_S, I_{DSS})$

- Q-point: the intersection point between the DC load line and the transfer curve.

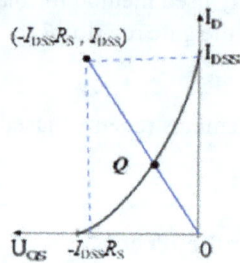

Figure 7.17   Q-point in self-biasing

Example: Plot the DC load line and determine the Q-point for the circuit shown in Figures 7.16 and the transfer characteristics curve shown in Figures 7.18, if R_S is 600 Ω, and I_{DSS} is 8 mA.

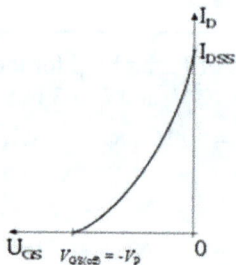

Figure 7.18   Transfer characteristic of JFET

- Given: $R_S = 600\ \Omega$, and $I_{DSS} = 8$ mA.
- Find: Plot DC-load line and determine the Q-point.
- Solution:
  - The first point: $(V_{GS},\ I_D) = (0,\ 0)$

    If $I_D = 0$, $V_{GS} = -I_D R_S = -(0\text{ mA})(600\ \Omega) = 0$ V

  - The second point: $(V_{GS},\ I_D) = (-I_{DSS}R_S,\ I_{DSS})$

    $V_{GS} = -I_D R_S = -I_{DSS}R_S = -(8\text{ mA})(0.6\text{ k}\Omega) = -4.8$ V

    $(V_{GS},\ I_D) = (-4.8\text{ V},\ 8\text{ mA})$

  - Connect the two points → DC load line

    $$(0,\ 0),\ (-4.8\text{ V},\ 8\text{ mA})$$

  - Q-point: the intersection point between the DC load line and the transfer curve.

    $(V_{GS},\ I_D) \approx (-2\text{ V},\ 3\text{ mA})$

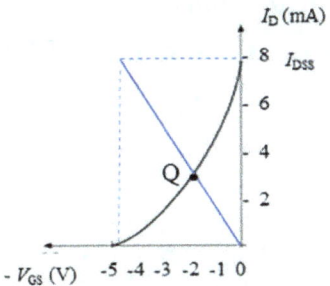

Figure 7.19   Q-point of the example

Midpoint biasing

- Midpoint biasing: a JFET with a near centered Q-point on transfer characteristic curve.
- Midpoint biasing represents the most efficient use of the device range (get the maximum amount of drain current I_D).
- The midpoint bias occurs when $V_{GS} \approx 0.3 \ V_{GS(off)}$ and drain current $I_D \approx 0.5 \ I_{DSS}$.

When $V_{GS} \approx 0.3 \ V_{GS(off)} \rightarrow I_D \approx 0.5 \ I_{DSS}$ (midpoint)

Derive: $I_D = I_{DSS} \left(1 - \dfrac{V_{GS}}{V_{GS(off)}} \right)^2 = I_{DSS} \left(1 - \dfrac{0.3 \ V_{GS(off)}}{V_{GS(off)}} \right)^2 = 0.49 \ I_{DSS} \approx 0.5 \ I_{DSS}$

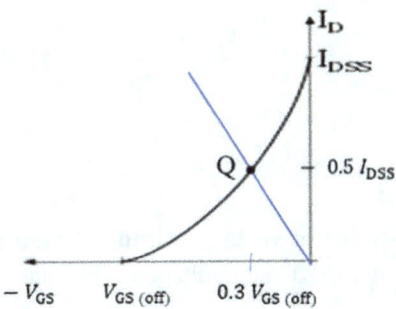

Figure 7.20   Midpoint biasing

7.2.3   JFET Voltage Divider-Bias

Voltage divider-bias circuit (N-channel JFET):

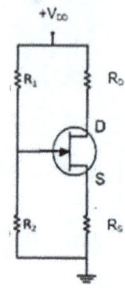

$+V_{DD}$

R_1 R_D

D

S

R_2 R_S

Figure 7.21   Voltage-divider bias

Stability of a JFET voltage-divider bias

- Voltage-divider bias is one of the most frequently used JFET biasing circuits.
- It uses a voltage divider circuit to provide good Q-point stability.
- The voltage-divider bias circuit of a JFET is very similar to the voltage-divider bias circuit of a BJT, but the DC analysis of each is different.

Calculating DC voltages and currents

- Gate-source voltage V_{GS}: $V_{GS} = V_G - V_S$

$$V_G \approx V_{DD}\,\frac{R_2}{R_1 + R_2} \qquad V_S = I_D R_S \qquad I_S \approx I_D$$

- Drain-source voltages V_{DS}: $V_{DS} = V_{DD} - I_D(R_D + R_S)$ $V_{DD} = I_D R_D + V_{DS} + I_S R_S$

- Drain current I_D: $I_D = \dfrac{V_{DD} - V_D}{R_D}$ $V_{DD} = V_D + I_D R_D$

  or $I_D = \dfrac{V_G - V_{GS}}{R_S}$ $V_G = V_{GS} + I_D R_S,\ \ I_D \approx I_S$

DC load line and Q-point in voltage-divider biasing

- Plot DC load line for a JFET with voltage-divide bias:

  – The first point: when $V_{GS} = 0$, $I_D = \dfrac{V_G}{R_S}$ $(V_{GS}, I_D) = (0, \dfrac{V_G}{R_S})$

  $$\text{Derive: } I_D = \frac{V_G - V_{GS}}{R_S} = \frac{V_G - 0}{R_S} = \frac{V_G}{R_S}$$

  – The second point: when $I_D = 0$, $V_{GS} = V_G$ $(V_{GS}, I_D) = (V_G, 0)$

  Derive: $V_{GS} = V_G - V_S = V_G - I_D R_S = V_G - (0)(R_S) = V_G$

- Plot these two points on the transfer characteristics curve and connect them by a line segment – DC load line.
- Q-point: the intersection point between the DC load line and the transfer curve.

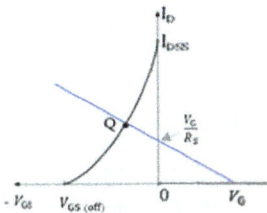

Figure 7.22   Q-point in voltage-divider biasing

Example: Determine the values of V_D, V_S, V_{GS}, and V_{DS} for the circuit shown in Figure 7.21, if R_1 is 3 MΩ, R_2 is 1.5 MΩ, R_D is 2 kΩ, R_S is 1.5 kΩ, V_{DD} is 12V, and I_D is 3 mA. Also plot the DC load line and determine the Q-point.

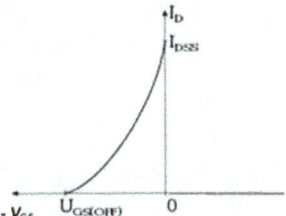

Figure 7.23   Transfer characteristic

- Given: $R_1 = 3$ MΩ, $R_2 = 1.5$ MΩ, $R_D = 2$ kΩ, $R_S = 1.5$ kΩ, $V_{DD} = 12$V, and $I_D = 3$ mA.

- Find: a) V_D, V_S, V_{GS}, and V_{DS}.
  b)   Plot the DC-load line and determine the Q-point.

- Solution: a)

  - V_D: $V_D = V_{DD} - I_D R_D = 12$ V $- (3$ mA$)(2$ kΩ$) = 6$ V   Milli: 10^{-3}; Kilo: 10^3

  - V_S: $V_S = I_D R_S = (3$ mA$)(1.5$ kΩ$) = 4.5$ V

  - V_{GS}:   Thinking process: $V_{GS} = V_G - V_S \rightarrow V_G = V_{DD}\dfrac{R_2}{R_1 + R_2}$

    $$V_G \approx V_{DD}\,\frac{R_2}{R_1 + R_2} = 12\text{ V}\,\frac{1.5 \text{ MΩ}}{3 \text{ MΩ} + 1.5 \text{ MΩ}}\quad 4\text{ V}$$

    $V_{GS} = V_G - V_S = 4$ V $- 4.5$ V $= -0.5$ V

  - V_{DS}: $V_{DS} = V_{DD} - I_D(R_D + R_S) = 12$ V $- (3$ mA$)(2$ kΩ $+ 1.5$ kΩ$)$
    $= 1.5$ V

- Solution: b)
    - The first point: $(V_{GS}, I_D) = (0, \dfrac{V_G}{R_S})$

$$I_D = \frac{V_G}{R_S} = \frac{4\ V}{1.5\ k\Omega} \approx 2.67\ mA$$

$$(0, \frac{V_G}{R_S}) = (0,\ 2.67\ mA)$$

    - The second point: $(V_{GS}, I_D) = (V_G, 0) = (4\ V,\ 0)$
      Connect the two points → DC load line
      $(0, 2.67\ mA),\quad (4V, 0)$

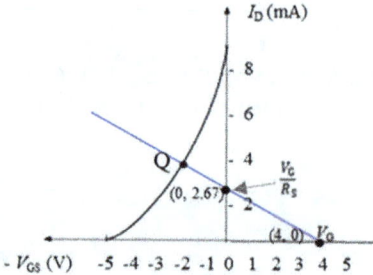

Figure 7.24   Determine the Q point

    - Q point: the intersection point between the DC load line and the transfer curve. $(V_{GS}, I_D) \approx (-2\ V, 4\ mA)$

7.3   MOSFET (Metal–Oxide–Semiconductor Field-Effect Transistor)

7.3.1   Introduction to MOSFET

Characteristics of MOSFETs

- There are two major types of FET: JFET (Junction gate field-effect transistor) and MOSFET (metal–oxide–semiconductor field-effect transistor).
- JFET: a semiconductor device with a PN junction that operates in only depletion mode.
- MOSFET: a special type of FET without a PN junction and operates in both depletion and enhancement mode.
    - The gate of MOSFET is insulated from the channel by a thin SiO_2 (silicon dioxide) layer that can reduce the gate leakage current.
    - MOSFET has a higher input impedance than JFET that can reduce the input current required and therefore the heat generated. MOSFET is the advanced form of JFET.

– The MOSFET's thermal stability and other characteristics make it very popular in the design and construction of integrated circuits for digital computers.

Table 7.3   The main difference between JFET and MOSFET

MOSFET (metal–oxide–semiconductor field-effect transistor)	JFET (junction gate field-effect transistor)
Operates in both depletion and enhancement mode	Operates in only depletion mode
Higher input impedance than JFET	High input impedance
The gate is insulated from the channel by a thin SiO$_2$ (silicon dioxide) layer.	The gate and channel are separated by a PN junction
Four types: N-channel depletion N-channel enhancement P-channel depletion P-channel enhancement	Two types: N-channel P-channel

- Basic structure of a N-channel MOSFET:

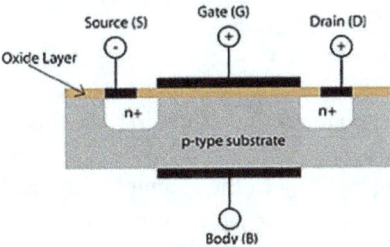

Figure 7.25   N-channel MOSFET

Types of MOSFETs

- Depletion and enhancement: there are two major types of MOSFET – depletion and enhancement.
  - Enhancement-mode MOSFET (E-MOSFET) with positive gate bias voltage (V_{GG}) to increase the channel width and switch the device "on" (switch on).
  - Depletion-mode MOSFET (D-MOSFET) with either positive or negative gate bias voltage (V_{GG}) to reduce the channel width (deplete or destroy the channel) and switch the device "off" (switch off).
- N- or P-channel: a MOSFET consists of a channel of N- or P-type semiconductor material through which current flows.
- Four types of MOSFETs: a depletion mode MOSFET and an enhancement-mode MOSFET are further classified as N-channel and P-channel.
  - N-channel depletion
  - N-channel enhancement

- P-channel depletion
- P-channel enhancement

Depletion-mode MOSFET

- Symbols of D-MOSFET:

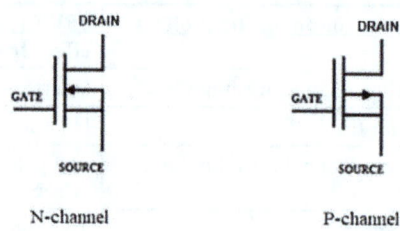

Figure 7.26   Symbols of D-MOSFET

- N-channel D-MOSFET: the arrow is pointing in.
- P-channel D-MOSFET: the arrow is pointing out.
- Basic operation: by applying a reverse bias (RB) voltage V_{GS} to the gate-source junction can reduce the width of the channel, increase the resistance of the channel, and reduce the drain current I_D.
  Apply a $V_{GS} \rightarrow$ gate-souse RB $\rightarrow$ channel width $\downarrow \rightarrow$ channel $R \uparrow \rightarrow I_D \downarrow \rightarrow$ switch off
- JFETs and MOSFETs are similar in their operating principles and the main difference is:
  - In JFETs, the conductivity of the channel is controlled by the electric field across the reverse-biased gate-souse junction.
  - In MOSFET, the conductivity of the channel is controlled by a transverse electric field across the insulating layer (SiO_2) embedded on the semiconductor.

    Transverse electric field, electromagnetic waves are perpendicular to the direction of travel.

Enhancement-mode MOSFET

- Symbols of E-MOSFET:

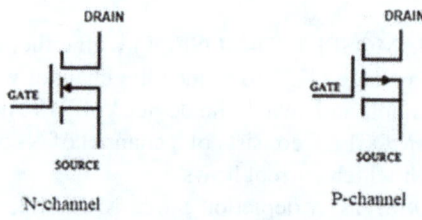

Figure 7.27   Symbols of E-MOSFET

- Basic operation: by applying a forward bias (FB) voltage V_{GS} to the gate-source junction can increase the width of the channel, reduce the resistance of the channel, and increase the drain current I_D.
  Apply a $V_{GS} \rightarrow$ gate-souse FB $\rightarrow$ channel width $\uparrow \rightarrow$ channel $R \downarrow \rightarrow I_D \uparrow \rightarrow$ switch on

7.3.2   MOSFET Transfer Characteristics

Transfer characteristic of D-MOSFET (N-channel)

- Transfer characteristic (or transconductance curve): the curve plot between the gate-source voltage V_{GS} and the drain current I_D.   A plot of V_{GS} verses I_D.
- The transfer characteristic of D-MOSFET is similar to the JFET.
  - When V_{GS} is zero, the maximum drain current I_D flowing is I_{DSS}.

    When $V_{GS} = 0$: $I_D = I_{DSS}$

  - When $V_{GS} = -V_P$, the transfer curve develops from $I_D = I_{DSS}$ at $V_{GS} = 0$, to $I_D = 0$ at $V_{GS} = -V_P$ (N-channel).

When $V_{GS} = -V_P : I_D = 0$
V_p: pinch-off voltage

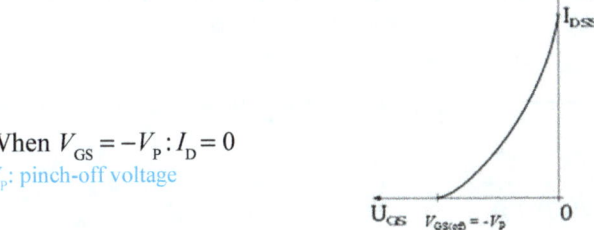

Figure 7.28   Transfer characteristic of D-MOSFET

- Shockley's equation: the transfer characteristic of MOSFET is defined by Shockley's equation (to calculate drain current I_D or plot a transfer curve).

$$I_D = I_{DSS} \left(1 - \frac{V_{GS}}{V_{GS(off)}} \right)^2$$

Transfer characteristic of E-MOSFET (N-channel)

- Recall: E-MOSFET with positive gate bias voltage ($+V_{GG}$) to increase the channel width and switch the device "ON".
- When V_{GS} is zero, the drain current I_D is zero.
  When $V_{GS} = 0$: $I_D = 0$

- When V_{GS} is equal or great than the threshold voltage $V_{GS(Th)}$, the channel width increases, and the drain current I_D increases.

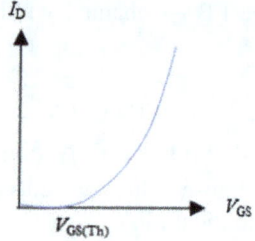

Figure 7.29   Transfer characteristic of E-MOSFET

When $V_{GS} \geq V_{GS(Th)} : I_D \uparrow$

- $V_{GS\ (Th)}$ (threshold voltage): the minimum gain-source voltage required to produce I_D.
- The $V_{GS\ (Th)}$ of the N-channel E-MOSFET is similar to the V_{BE} of the BJT.

- The equation for the transfer characteristic of E-MOSFET is given as:

$$I_D = k\,(V_{GS} - V_{GS(Th)})^2 \qquad k - \text{a constant for the MOSFET}$$

$$k = \frac{I_{D(on)}}{(V_{GS} - V_{GS(Th)})^2} \qquad V_{GS(Th)} \text{ and } I_{D(on)} \text{ can be obtained from the datasheet.}$$

7.3.3   E-MOSFET Biasing

The commonly used methods of E-MOSFET biasing

- Drain feedback bias
- Voltage-divider bias

Drain-feedback bias of E-MOSFET

- Drain-feedback bias circuit for a E-MOSFET (N-channel):

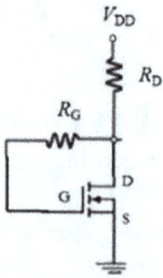

Figure 7.30   Drain-feedback bias

- Drain-feedback bias for an E-MOSFET is very similar to a collector-feedback bias circuit of a BJT.
- The gate resistor R_G of the collector-feedback bias circuit is connected to the drain D rather than to V_{DD}.

Calculating DC voltages and currents

- The gate circuit I_G ($I_G \approx 0$) is negligible since the gate has a very high impedance. Therefore, the drain and gate voltages are the same ($V_G = V_D$) by the gate resistor ($V_{RG} = 0$).
- The drain-source voltage V_{DS}: $V_{DS} \approx V_D = V_{DD} - I_D R_D$

- The drain current I_D:

$$I_D \approx I_S = \frac{V_{DD} - V_{DS}}{R_D}$$

- The gate-source voltage V_{GS}: $V_{GS} = V_{DS}$ $V_G = V_D$

Example: Determine the values of V_{GS} for the circuit shown in Figure 7.30, if R_D is 2 kΩ, R_G is 3 MΩ, V_{DD} is 12V, and I_D is 2 mA.
- Given: $R_D = 2\,\text{k}\Omega$, $R_G = 3\,\text{M}\Omega$, $V_{DD} = 12\,\text{V}$, $I_D = 2\,\text{mA}$.
- Find: V_{GS}
- Solution:

V_{GS}: $V_{GS} = V_{DS}$?
$V_{DS} = V_{DD} - I_D R_D = 12\,\text{V} - (2\,\text{mA})(2\,\text{k}\Omega) = 8\,\text{V}$

Milli: 10^{-3}; Kilo: 10^3

$V_{GS} = V_{DS} = 8\,\text{V}$

Voltage-divider bias of E-MOSFET biasing

- Voltage-divider bias circuit for a E-MOSFET (N-channel):

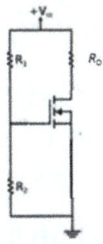

Figure 7.31   Voltage-divider bias

- Voltage-divider bias circuit for E-MOSFET is very similar to the voltage-divider bias circuit of a BJT or JFET, but the DC analysis of each is different.

Calculating DC voltages and currents

- The gate-source voltage V_{GS}:
$$V_{GS} \approx V_{DD} \frac{R_2}{R_1 + R_2}$$

- The drain-source voltage V_{DS}:
$$V_{DS} = V_{DD} - I_D R_D$$

- The drain current I_D:
$$I_D = k (V_{GS} - V_{GS(Th)})^2$$

$$k = \frac{I_{D(on)}}{(V_{GS} - V_{GS(Th)})^2}$$

Example: Determine the values of I_D and V_{DS} for the circuit shown in Figure 7.31, if R_1 is 3 MΩ, R_2 is 1.5 MΩ, R_D is 100 Ω, V_{DD} is 12 V, $V_{GS(Th)}$ is 2V, and $I_{D(On)}$ is 5 mA.
- Given: $R_1 = 3$ MΩ, $R_2 = 1.5$ MΩ, $R_D = 100$ Ω, $V_{DD} = 12$ V, $V_{GS(Th)} = 2$V, $I_{D(on)} = 100$ mA
- Find: I_D and V_{DS}.
- Solution:
  - I_D: Thinking process: $I_D = k (V_{GS} - V_{GS(Th)})^2$ $\rightarrow V_{GS} \approx V_{DD} \frac{R_2}{R_1 + R_2}$

$$\rightarrow k = \frac{I_{D(on)}}{(V_{GS} - V_{GS(Th)})^2}$$

$$V_{GS} \approx V_{DD} \frac{R_2}{R_1 + R_2} \quad 12 \text{ V} \frac{1.5 \text{ MΩ}}{3 \text{ MΩ} + 1.5 \text{ MΩ}} = 4 \text{ V}$$

$$k = \frac{I_{D(on)}}{(V_{GS} - V_{GS(Th)})^2} = \frac{100 \text{ mA}}{(4 \text{ V} - 2 \text{ V})^2} = 25 \text{ mA/ V}^2$$

$$I_D = k (V_{GS} - V_{GS(Th)})^2 = (25 \text{ mA/V}^2)(4\text{V} - 2 \text{ V})^2 = 100 \text{ mA}$$

  - V_{DS}: $V_{DS} = V_{DD} - I_D R_D = 12 \text{ V} - (100 \text{ mA})(0.1 \text{ kΩ}) = 2 \text{ V}$
        Milli: 10^{-3}; Kilo: 10^3

7.3.4   D-MOSFET Biasing

The commonly used methods of D-MOSFET biasing

- The D-MOSFET can be biased using some methods used with the JFET (gate bias, self- bias, voltage-divider bias, etc.).
- A simple bias method for the D-MOSFET is zero bias (set $V_{GS} = 0$).

Zero bias of D-MOSFET

- Zero bias circuit for a D-MOSFET (N-channel):

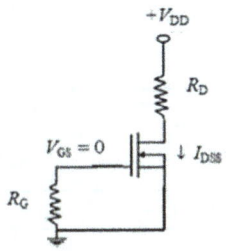

Figure 7.32   Zero bias

- Zero bias is so named because it operates at V_{GS} at 0 V. Similar to self-bias, it does not require a second DC source for the gate terminal.

Calculating DC voltages and currents

- The gate-source voltage V_{GS}: $V_{GS} = 0$ Zero bias
- The drain current I_D: $I_D = I_{DSS}$ $V_{GS} = 0$
- The drain-source voltage V_{DS}: $V_{DS} = V_{DD} - I_{DSS} R_D$ $I_D = I_{DSS}$

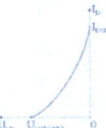

Figure 7.33   Transfer characteristic

Example: Determine the values of drain current for the circuit shown in Figure 7.32, if V_{DD} is 15 V, R_D is 1.5 kΩ, R_G is 5 MΩ, and V_{DS} is 6 V.

- Given: $V_{DD} = 15$ V, $R_D = 1.5$ kΩ, $R_G = 5$ MΩ, $V_{DS} = 6$ V
- Find: I_D
- Solution: Thinking process: $I_D = I_{DSS} \rightarrow V_{DS} = V_{DD} - I_{DSS} R_D$

$$I_{DSS} = \frac{V_{DD} - V_{DS}}{R_D} = \frac{15\ V - 6\ V}{1.5\ kΩ} = 6\ mA \qquad \text{Milli: } 10^{-3}; \text{ Kilo: } 10^3$$

Summary

Types of FET

- JFET (junction gate field-effect transistor): a semiconductor device with a PN junction that operates in only depletion mode.

- MOSFET (metal–oxide–semiconductor field-effect transistor): a semiconductor device without a PN junction that operates in both depletion and enhancement mode.

Table 7.4:   The main difference between FET and BJT

FET (Field-effect transistor)	BJT (bipolar junction transistor)
Only the majority charge carriers (holes or electrons) flow	Both majority and minority charge carriers flow
A unipolar device (only one carrier type)	A bipolar device
A voltage-control device	A current-controlled device
High input resistance	Low input resistance
Low power consumption	High power consumption
Low noise level	High noise level

Basic structure of JFET

- N-channel JFET: a JFET whose channel is composed of a majority of electrons as the charge carrier. It made up a bar of N-type semiconductor with two parts of P-type material inserted in the sides.
- P-channel JFET: a JFET whose channel is composed of a majority of holes as the charge carrier. It made up a bar of P-type material with two parts of N-type material inserted in the sides.

JFET – a voltage-controlled device

- Reverse bias voltage V_{GS}: by applying a reverse bias voltage V_{GS} to the gate-source junction can control the width of the channel and the flow of current from the source to the drain.
- Voltage-controlled device:
  $V_{GS} \rightarrow$ depletion layer $\rightarrow$ channel width $\rightarrow$ resistance of the channel $\rightarrow I_D$

The V_{GS} of a JFET

- A N-channel JFET requires a negative V_{GS}.
- A P-channel JFET requires a positive V_{GS}.

JFET characteristics curve

- Drain (or output) characteristic curve: A plot of V_{DS} verses I_D.
- JFET drain curve vs. BJT collector curve:

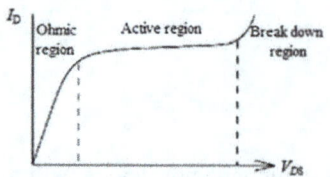

Figure 7.34(a)   JFET drain curve

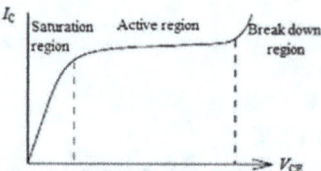

Figure 7.34(b)   BJT collector curve

Pinch-off voltage vs. cut-off voltage

- Pinch-off voltage V_{P}: the minimum value of V_{DS} at which I_{D} becomes saturation ($I_{\text{D}} = I_{\text{DSS}}$) at zero V_{GS}.
- Cut-off voltage $V_{\text{GS(off)}}$: the minimum value of V_{GS} at which I_{D} becomes approximately zero.
- $V_{\text{GS(off)}} = -V_{\text{P}}$

Transfer characteristic of JFET (N-channel)

- Transfer characteristic (or transconductance curve): a plot of V_{GS} versus I_{D}.

- Shockley's equation: $I_{\text{D}} = I_{\text{DSS}} \left(1 - \dfrac{V_{\text{GS}}}{V_{\text{GS(off)}}} \right)^2$

Midpoint biasing

- Q-point of JFET: V_{GS} and I_{D}
- Midpoint biasing: a JFET with a near centered Q-point on transfer characteristic curve. It represents the most efficient use of the device range.
- The midpoint bias occurs when $V_{\text{GS}} \approx 0.3\,V_{\text{GS(off)}}$ and $I_{\text{D}} \approx 0.5\,I_{\text{DSS}}$.

Table 7.5: Methods of JFET Biasing

Biasing	Self-bias	Voltage-divider bias	Gate bias (or fixed bias)
Circuit			
V_{GS}	$V_{\text{GS}} = -I_{\text{D}}R_{\text{S}}$	$V_{\text{GS}} = V_{\text{G}} - V_{\text{S}}$ $V_{\text{G}} \approx V_{\text{DD}} \dfrac{R_2}{R_1 + R_2}$ $V_{\text{S}} = I_{\text{D}} R_{\text{S}}$	$V_{\text{GS}} = -V_{\text{GG}}$
V_{DS}	$V_{\text{DS}} = V_{\text{DD}} - I_{\text{D}}(R_{\text{D}} + R_{\text{S}})$	$V_{\text{DS}} = V_{\text{DD}} - I_{\text{D}}(R_{\text{D}} + R_{\text{S}})$	$V_{\text{DS}} = V_{\text{DD}} - I_{\text{D}}R_{\text{D}}$
I_{D}	$I_{\text{D}} \approx I_{\text{S}}$	$I_{\text{D}} = \dfrac{V_{\text{DD}} - V_{\text{D}}}{R_{\text{D}}}$ or $I_{\text{D}} = \dfrac{V_{\text{G}} - V_{\text{GS}}}{R_{\text{S}}}$	$I_{\text{D}} = I_{\text{DSS}} \left(1 - \dfrac{V_{\text{GS}}}{V_{\text{GS(off)}}} \right)^2$

Characteristics of MOSFETs

- JFET (junction gate field-effect transistor): a semiconductor device with a PN junction that operates in only depletion mode.
- MOSFET(metal–oxide–semiconductor field-effect transistor): a special type of FET without a PN junction that operates in both depletion and enhancement mode.

Table 7.6:   The main difference between JFET and MOSFET

MOSFET (metal–oxide–semiconductor field-effect transistor)	JFET (junction gate field-effect transistor)
Operates in both depletion and enhancement mode	Operates in only depletion mode
Higher input impedance than JFET	High input impedance
The gate is insulated from the channel by a thin SiO_2 (silicon dioxide) layer	The gate and channel are separated by a PN junction

Four types: N-channel depletion
            N-channel enhancement
            P-channel depletion
            P-channel enhancement
	Two types: N-channel
	            P-channel

Types of MOSFETs

- Enhancement-mode MOSFET (E-MOSFET) with positive gate bias voltage (V_{GG}) to increase the channel width and switch the device "on" (switch on).
- Depletion-mode MOSFET (D-MOSFET) with either positive or negative gate bias voltage (V_{GG}) to reduce the channel width and switch the device "off" (switch off).

Depletion-mode MOSFET

- Basic operation:
  Apply a V_{GS} → gate-souse RB → channel width ↓→ channel R ↑→ I_D ↓→ switch off
- JFETs and MOSFETs are similar in their operating principles and the main difference is:
  - In JFETs, the conductivity of the channel is controlled by the electric field across the reverse-biased gate-souse junction.
  - In MOSFET, the conductivity of the channel is controlled by a transverse electric field across the insulating layer (SiO_2) embedded on the semiconductor.

Enhancement-mode MOSFET

- Basic operation:
  Apply a V_{GS} → gate-souse FB → channel width ↑→ channel R ↓→ I_D ↑→ switch on

Transfer characteristic of D-MOSFET (N-channel)

- Transfer characteristic (or transconductance curve): a plot of V_{GS} verses I_D.
- When $V_{GS} = 0$: $I_D = I_{DSS}$
- When $V_{GS} = -V_P$: $I_D = 0$
- Shockley's equation: the transfer characteristic of MOSFET is defined by

  Shockley's equation. $I_D = I_{DSS} \left(1 - \dfrac{V_{GS}}{V_{GS(off)}}\right)^2$

Transfer characteristic of E-MOSFET (N-channel)

- When $V_{GS} = 0$: $I_D = 0$
- When $V_{GS} \geq V_{GS(Th)}$: $I_D = \uparrow$
- The equation for the transfer characteristic of E-MOSFETs is given as:

$$I_D = k\,(V_{GS} - V_{GS(Th)})^2 \qquad k = \frac{I_{D(on)}}{(V_{GS} - V_{GS(Th)})^2}$$

The commonly used methods of E-MOSFET biasing

- Drain feedback bias
- Voltage-divider bias

The commonly used methods of D-MOSFET biasing

- The DE-MOSFET can be biased using some methods used with the JFET (gate bias, self-bias, voltage-divider bias, etc.).
- A simple bias method for the DE-MOSFET is zero bias (set $V_{GS} = 0$).

Table 7.7: Methods of MOSFET Biasing

Biasing	Drain-feedback bias	Voltage-divider bias	Zero bias of D-MOSFET
Circuit			
V_{GS}	$V_{GS} = V_{DS}$	$V_{GS} = V_{DD}\dfrac{R_2}{R_1 + R_2}$	$V_{GS} = 0$
V_{DS}	$V_{DS} \approx V_D = V_{DD} - I_D R_D$	$V_{DS} = V_{DD} - I_D R_D$	$V_{DS} = V_{DD} - I_{DSS} R_D$
I_D	$I_D \approx I_S = \dfrac{V_{DD} - V_{DS}}{R_D}$	$I_D = k\,(V_{GS} - V_{GS(Th)})^2$ $\qquad k = \dfrac{I_{D(on)}}{(V_{GS} - V_{GS(Th)})^2}$	$I_D = I_{DSS}$

Table 7.8:   Comparison of the JFET and BJT

	FET (Field-effect transistor)	BJT (bipolar junction transistor)
Symbol		
Terminal	D – drain G – gate S – source	C – collector B – base E – emitter
DC voltages	 V_{DD} and $-V_{GG}$ (For a N-channel JFET)	 V_{CC} and V_{BB} (For a NPN BJT)
Characteristic curves		
Four regions	 – Ohmic region – Active region (or satu-ration region) – Breakdown region – Cut-off region (or pinch-off region)	 – Saturation region – Active region – Breakdown region – Cut-off region
V- or *I*-controlled device	A voltage-controlled device: V_{GS} controls I_D	A current-controlled device: I_B controls I_C

Types of Transistors

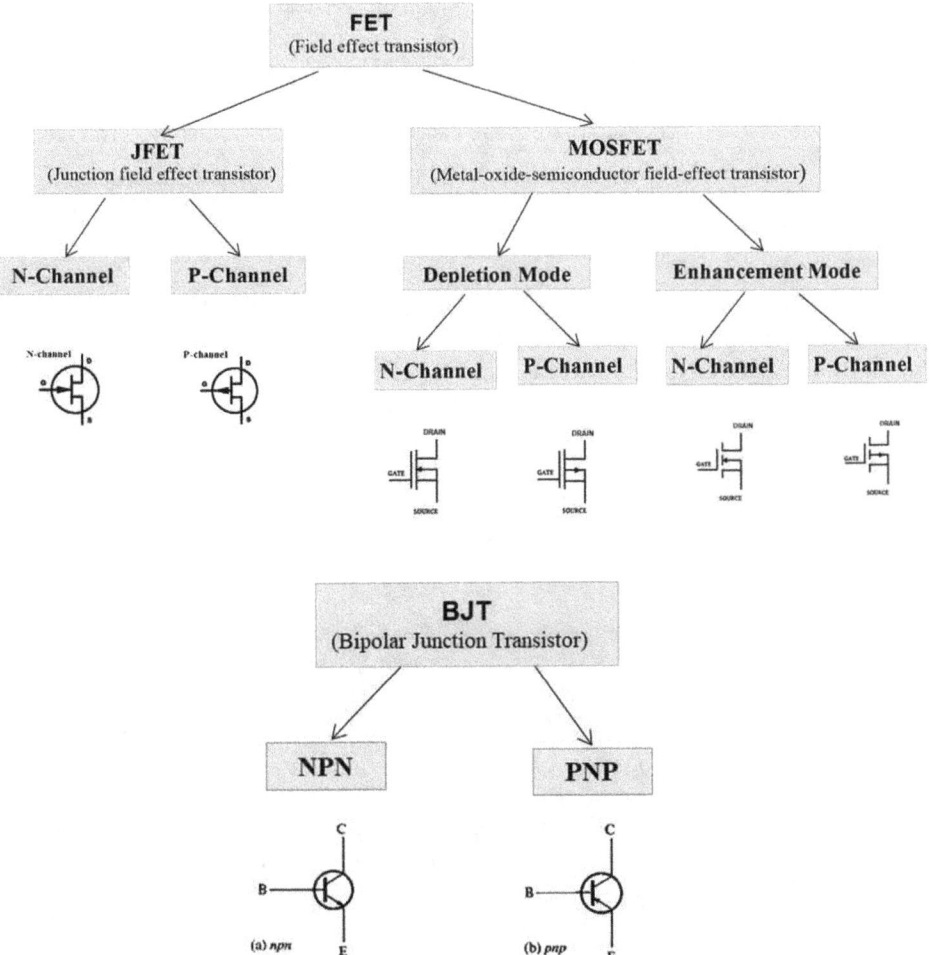

Self-Test

7.1 1. The field-effect transistor (FET) is a (   )-controlled device.

2. JFET is a semiconductor device with a PN junction that operates in only (   ) mode.

3. The P-channel JFET is a JFET whose channel is composed of a majority of (   ) as the charge carrier.

4. The (   ) is used to control current flow between the other two terminals.

5. Since a JFET must be operated such that the gate-source junction is always reverse-biased. This requires a (   ) V_{GS} for an N-channel JFET.

6. The (   ) voltage is the minimum value V_{DS} at which maximum drain current I_D can flow.

7. The (   ) characteristic is the curve plot between gate-source voltage V_{GS} and drain current I_D.

8. Plot the transfer characteristics curve, if I_{DSS} is 2 mA and $V_{GS(off)}$ is negative 5 V.

7.2 9. The (   ) current is determined by the transfer characteristic or the Shockley's equation of JFET.

10. Determine the values of V_{GS} and V_{DS} for the circuit shown in Figure 7.35, if R_D is 3 kΩ, R_S is 400 Ω, V_{DD} is 10 V, and I_D is 2 mA.

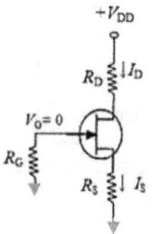

Figure 7.35   Ch 7: No. 10, self-test

11. Plot the DC load line and determine the Q-point for the circuit and the transfer characteristics curve shown in Figure 7.36(a) and (b), if R_D is 1 kΩ, R_S is 500 Ω, and I_{DSS} is 6 mA.

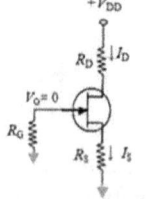

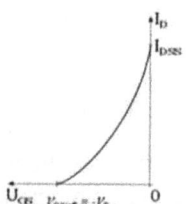

Figure 7.36(a)   Ch 7: No. 11, self-test   Figure 7.36(b)   Ch 7: No. 11, self-test

12. The midpoint bias occurs when $V_{GS} \approx 0.3 V_{GS(off)}$ and drain current $I_D \approx (\quad) I_{DSS}$.

13. (   ) bias is one of the most frequently used JFET biasing circuits.

14. **(a)** Determine the values of V_D, V_S, V_{GS}, and V_{DS} for the circuit shown in Figure 7.37, if R_1 is 4 MΩ, R_2 is 2 MΩ, R_D is 1 kΩ, R_S is 2.5 kΩ, V_{DD} is 15 V, and I_D is 4 mA. **(b)** Plot the DC load line and determine the Q-point.

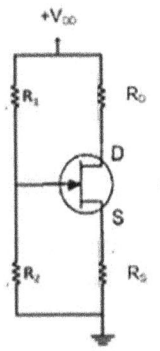

Figure 7.37   Ch 7: No. 14, self-test

7.3 **15.** MOSFET is a special type of FET without a PN junction and operates in both depletion and (   ) mode.

16. (   )-mode MOSFET with either positive or negative gate bias voltage to reduce the channel width and switch the device "off".

17. In MOSFET, the conductivity of the channel is controlled by a transverse electric field across the (   ) layer embedded on the semiconductor.

18. Drain-feedback bias for E-MOSFET is very similar to a (   )-feedback bias circuit of BJT.

19. Determine the values of V_{GS} for the circuit shown in Figure 7.38, if R_D is 3 kΩ, R_G is 4 MΩ, V_{DD} is 15V, and I_D is 3 mA.

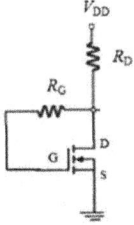

Figure 7.38   Ch 7: No. 19, self-test

20. Determine the values of I_D and V_{DS} for the circuit shown in Figure 7.39, if R_1 is 4 MΩ, R_2 is 2 MΩ, R_D is 120 Ω, V_{DD} is 15 V, $V_{GS(Th)}$ is 3 V, and $I_{D(on)}$ is 110 mA.

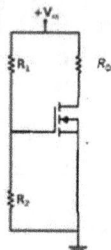

Figure 7.39   Ch 7: No. 20, self-test

21. Determine the values of drain current for the circuit shown in Figure 7.40, if V_{DD} is 12 V, R_D is 1 kΩ, and V_{DS} is 6 V.

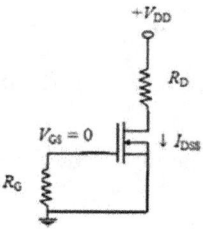

Figure 7.40   Ch 7: No. 21, self-test

Chapter 8

AC analysis of FET circuits – FET amplifiers

Chapter outline

8.1   JFET equivalent model

8.1.1   Introduction to FET amplifier

FET amplifier

- Recall – amplifier: an electronic device that increases the voltage, current, or power of an input signal. It is a circuit that has a power gain greater than 1.
- Recall – amplifier gain: the ratio of output to input in the form of power, voltage, and current. It is a measure of the ability of an amplifier.

$$A_\text{P} = \frac{P_\text{out}}{P_\text{in}} \qquad A_\text{V} = \frac{V_\text{out}}{V_\text{in}} \qquad A_\text{i} = \frac{i_\text{out}}{i_\text{in}}$$

- FET amplifier: an FET in a linear region around an operating point can work as an amplifier to amplify a small AC input signal (low-level signal such as a wireless signal).
- A small-signal JFET (N-channel) amplifier with voltage-divider bias:
  - Recall – the purpose of biasing is to stabilize the Q-Point (DC operation) in the desired region of operation in response to an AC input signal.
  - A small AC voltage (V_in) is applied to the input and passed to output (V_out) through coupling capacitors that are used to pass an AC signal without disturbing the Q-point of the circuit.

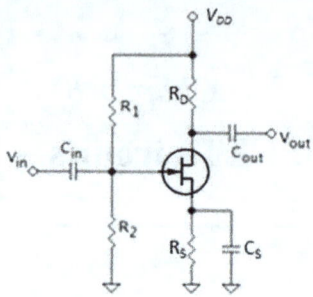

Figure 8.1   A JFET amplifier with voltage-divider bias

–   A bypass capacitor on the source resister (R_S) to bypass desirable signals to the ground and to increase gain.

Transconductance

*   Transconductance (g_m): the ratio of the change in current at the output to the change in voltage at the input.

    The "trans" is short for "transfer" (input to output).

    $$g_\text{m} = \frac{\text{change in } I_\text{out}}{\text{change in } V_\text{in}} \qquad g_\text{m} = y_\text{fs} \text{ (datasheets)}$$

*   Transconductance for FET (g_m): the ratio of the change in drain current I_d (output) to the change in gate voltage V_gs (input).

    $$AC\!: g_\text{m} = \frac{I_\text{d}}{V_\text{gs}}$$

    $$DC\!: g_\text{m} = \frac{\Delta I_\text{d}}{\Delta V_\text{gs}}$$

*   Definition of the transconductance g_m using transfer characteristics:

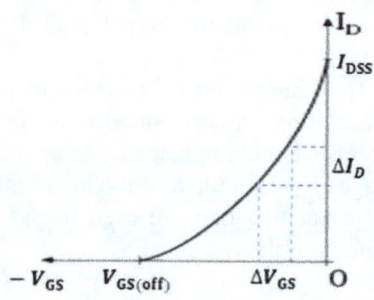

Figure 8.2   Transfer characteristics

8.1.2   JFET AC equivalent circuit

Build JFET AC equivalent circuit

* Small-signal model of JFET: an equivalent circuit that can be used to accurately predict the performance of a JFET. It can be used to estimate the input impedance, gain, and operating conditions of JFET amplifiers.
* JFET (N-channel) AC equivalent circuit:

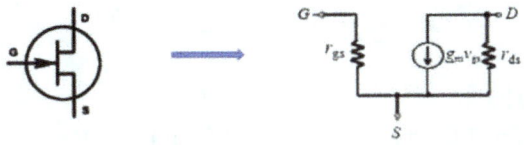

Figure 8.3   JFET AC equivalent circuit

FET internal AC resistance

* FET internal AC resistance: a small AC resistance looking into the terminal of an FET.
* Input resistance r_{gs}: the FET internal resistance between the gate and the source. Recall that an FET has a high input resistance, it is high enough to be approximated as an open circuit ($r_{gs} \approx \infty$).
* Output resistance r_{ds}: the FET internal resistance between the drain and the source. It is typically much larger (reverse-biased) than R_D and can be neglected in calculations (it can be replaced by an open circuit, $r_{ds} \approx \infty$).

Simplified JFET AC equivalent circuit:

* r_{gs} and r_{ds}: very large – open circuit
* The drain branch acts as a current source ($I_d = g_m v_{gs}$).

$$g_m = \frac{I_d}{V_{gs}}$$

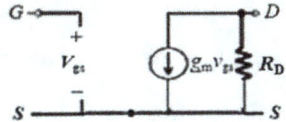

Figure 8.4   Simplified JFET equivalent circuit

JFET voltage gain

* Voltage gain without the load resistor R_L:

$$A_V = \frac{V_{out}}{V_{in}} = \frac{V_{ds}}{V_{gs}} = \frac{I_d R_D}{I_d / g_m} = g_m R_D$$

$$g_m = \frac{I_d}{V_{gs}}$$

* Voltage gain with the load resistor R_L:

$$A_V = \frac{V_{out}}{V_{in}} = g_m (R_D \| R_L)$$

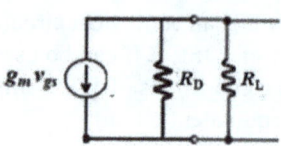

Figure 8.5   Voltage gain with R_L

The Q-point of JFET

- The Q-point of a JFET circuit is obtained from the values of the drain current I_D versus the drain-source voltages V_{DS} (for a specific V_{GS} on the DC load line).

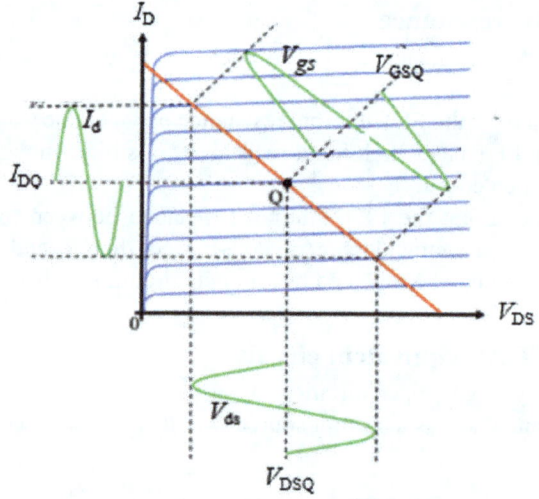

Figure 8.6   Midpoint biasing

Bias a JFET → establish a Q-point (I_D and V_{DS})

- Midpoint biasing: an amplifier with a centered Q-point on the DC load line can provide amplification of an AC input signal without distortion or clipping to the output waveform.

8.2   JFET amplifier analysis

8.2.1   Common-drain amplifier analysis

Common-drain (CD) circuit (source follower)

- Drain – common

- V_{in} – gate
- V_{out} – source

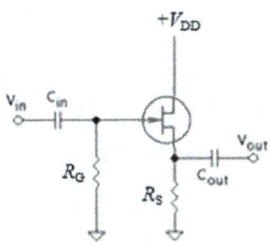

Figure 8.7   *Common-drain amplifier*

AC equivalent circuit

- AC equivalent circuit of a JFET common-drain (CD) amplifier is obtained by replacing all capacitors with a short circuit and setting DC voltage source V_{DD} to zero and replacing it with a ground.

The voltage gain of a CD amplifier

- Voltage gain without the load resistor R_L:

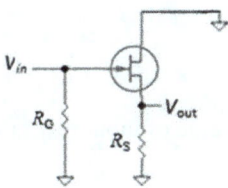

Figure 8.8   *AC equivalent circuit of a CD amplifier*

$$A_V = \frac{V_{out}}{V_{in}} = \frac{g_m R_s}{1 + g_m R_s}$$

Derive: $\quad A_V = \frac{V_{out}}{V_{in}} = \frac{I_d R_S}{V_{gs} + I_d R_S} = \frac{g_m V_{gs} R_S}{V_{gs} + g_m V_{gs} R_S} = \frac{g_m R_S}{1 + g_m R_S}$

$$g_m = \frac{I_d}{V_{gs}}, \quad I_d = g_m V_{gs}$$

- Voltage gain is approximately 1: if $g_m R_S \gg 1$, the voltage gain A_V is approximately equal to 1.

$$A_V \approx 1, \quad \text{if } g_m R_S \gg 1 \qquad A_V = \frac{g_m R_S}{1 + g_m R_S} \approx \frac{g_m R_S}{g_m R_S} = 1$$

- Voltage gain with the load resistor R_L:

$$A_V = \frac{V_{out}}{V_{in}} = \frac{g_m (R_S \| R_L)}{1 + g_m (R_S \| R_L)}$$

Phase relationship

- Source follower: the CD amplifier is also known as the source follower, because the load voltage "follows" the input signal very closely, and voltage gain is approximately equal to 1 ($A_v \approx 1$).
- In phase: CD amplifier receives its input signal to the gain with the output voltage taken from across the source load. As the source voltage follows the gain voltage, the output voltage in a CD amplifier is in phase with the input voltage (it simply follows it).

  The JFET's CD configuration (source follower) is similar to a BJT's common-collector configuration (voltage follower or emitter follower).

Input resistance

- The input resistance at the gate of the CD circuit (R_{ig}):

- $$R_{ig} = \frac{V_{GS}}{I_{GSS}}$$

  I_{GSS}: the gate-to-source leakage current (datasheet).

- The total circuit input resistance presented to the AC source of the CD circuit (R_{in}):

  $$R_{in} \approx R_G \| R_{ig} \approx R_G \qquad\qquad \text{If } R_{ig} \gg R_G$$

  - R_{ig} is typically much larger than RG and can be neglected in calculations.

  - I_{GSS} is leakage current and very small. $R_{ig} \uparrow\uparrow = \dfrac{V_{GS}}{I_{GSS}\downarrow\downarrow}$

- High input resistance: the value of input resistance on a CD amplifier is usually very high.

Output resistance

- The output resistance presented to the load of the CD circuit:

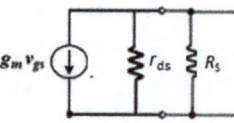

Figure 8.9   Output resistance

$$R_{out} \approx \frac{1}{g_m} \| R_S \qquad\qquad\qquad g_m = \frac{I_d}{V_{gs}}$$

$$R_{out} = \frac{1}{g_m} \| r_{ds} \| R_S \qquad\qquad \text{Conductance } G = \frac{1}{R}$$

$$\approx \frac{1}{g_m} \| R_S \qquad\qquad\qquad r_{ds} \approx \infty$$

- Low output resistance: the output resistance on a CD amplifier is always very low.

Example: For the JFET CD amplifier of Figure 8.7, determine the input and output resistance R_{in}, R_{out}, and voltage gain A_V, if R_G is 8 MΩ, R_S is 5 kΩ, R_L is 12 MΩ, V_{DD} is 12V, and g_m is 5 ms.

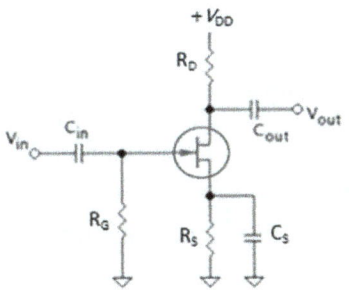

Figure 8.10   CS amplifier

- Given: $R_G = 8$ MΩ, $R_S = 5$ kΩ, $R_L = 12$ MΩ, $V_{DD} = 12$ V, and $g_m = 5$ ms.
- Find: R_{in}, R_{out}, and A_V.
- Solution:

R_{in}: $R_{in} \approx R_G = $ **8 MΩ**

R_{out}: $R_{out} \approx \dfrac{1}{g_m} \,\|\, R_S = \dfrac{1}{5\text{ ms}} \,\|\, 5\text{ kΩ}$ $\dfrac{1}{2\,\text{ms}} = \dfrac{1}{0.002\text{s}} = 500\ \Omega = 0.5\text{ kΩ; Kilo: }10^3$

$= 0.2$ kΩ $\backslash\backslash$ 5 kΩ $\approx$ **0.19 kΩ**

A_V: $A_V = \dfrac{g_m\,(R_S \,\|\, R_L)}{1 + g_m\,(R_S \,\|\, R_L)} = \dfrac{5\text{ ms }(5\text{ kΩ }\backslash\backslash 12\text{ MΩ})}{1 + 5\text{ ms }(5\text{ kΩ }\backslash\backslash 12\text{ MΩ})}$

$= \dfrac{5\text{ ms }(5\text{ kΩ }\backslash\backslash 12{,}000\text{ kΩ})}{1 + 5\text{ ms }(5\text{ kΩ }\backslash\backslash 12{,}000\text{ kΩ})} \approx \dfrac{5\text{ ms }(5\text{ kΩ})}{1 + 5\text{ ms }(5\text{ kΩ })} \approx$ **0.96** $A_V \approx 1$

8.2.2   Common-source amplifier analysis

Common-source (CS) circuit

- Source – common
- V_{in} – gate
- V_{out} – drain

AC equivalent circuit

- AC equivalent circuit of a JFET common-source (CS) amplifier is obtained by replacing all capacitors with a short circuit and setting DC voltage source V_{DD} to zero and replacing it with a ground.

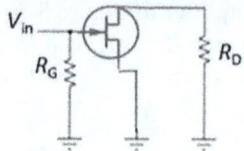

Figure 8.11   AC equivalent circuit of a CS amplifier

Input resistance

- The input resistance at the gate of the transistor:

$$R_{ig} = \frac{V_{GS}}{I_{GSS}}$$

I_{GSS}: the gate-to-source leakage current (datasheets).
- The total circuit input resistance presented to the AC source:
$$R_{in} = R_G \backslash\backslash R_{ig} \approx R_G, \quad \text{if } R_{ig} \gg R_G$$
R_{ig} is typically much larger than R_G and can be neglected in calculations.

Output resistance

- The output resistance presented to the load of the circuit:
$$R_{out} = R_D \backslash\backslash r_{ds}$$

- AC internal resistance between the drain and the source r_{ds} represents the reverse-biased drain-source junction resistance. It is typically much larger than R_D and can be replaced by an open circuit ($r_{ds} \approx \infty$).

$$R_{out} \approx R_D \qquad \text{If } r_{ds} \gg R_D$$

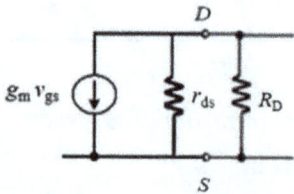

Figure 8.12   Output resistance

Voltage gain

- The voltage gain of JFET CS amplifier: the ratio of output to input in the form of voltage.
- Voltage gain without the load resistor R_L:

$$A_V = \frac{V_{out}}{I_{in}} = g_m R_D$$

Derive: $A_{\mathrm{V}} = \dfrac{V_{\mathrm{out}}}{V_{\mathrm{in}}} = \dfrac{V_{\mathrm{ds}}}{V_{\mathrm{gS}}} = \dfrac{I_{\mathrm{d}}R_{\mathrm{d}}}{I_{\mathrm{d}}/g_{\mathrm{m}}} = g_{\mathrm{m}}R_{\mathrm{D}}$ $g_{\mathrm{m}} = \dfrac{I_{\mathrm{d}}}{V_{\mathrm{gs}}}, V_{\mathrm{gs}} = \dfrac{I_{\mathrm{d}}}{g_{\mathrm{m}}}$

- Voltage gain with the load resistor R_{L}:

$$A_{\mathrm{V}} = \frac{V_{\mathrm{out}}}{V_{\mathrm{in}}} = g_{\mathrm{m}}\,(R_{\mathrm{D}} \parallel R_{\mathrm{L}})$$

Phase relationship
- 180° Phase shift: the output voltage in a CS amplifier is 180° out of phase with the input voltage.
- When the input signal increases, the drain current I_{d} also increases. An increase in drain current increases the voltage drop in the drain resistor R_{D}, reducing the voltage in the drain terminal V_{d} and producing a 180° phase shift.

$$V_{\mathrm{d}} = V_{\mathrm{DD}} - I_{\mathrm{d}}R_{\mathrm{D}}$$

Example: For the JFET CD amplifier of Figure 8.10, determine the input and output resistance R_{in}, R_{out}, and voltage gain A_{V}, if R_{G} is 5 MΩ, R_{S} is 500 Ω, R_{D} is 1.2 kΩ, R_{L} is 10 kΩ, V_{DD} is 15V, and g_{m} is 3 ms.
- Given: $R_{\mathrm{G}} = 5$ MΩ, $R_{\mathrm{S}} = 500$ Ω, $R_{\mathrm{D}} = 1.2$ kΩ, $R_{\mathrm{L}} = 10$ kΩ, $V_{\mathrm{DD}} = 15$V, and $g_{\mathrm{m}} = 3$ ms (Y_{fs} – datasheets).
- Find: R_{in}, R_{out}, and A_{V}.
- Solution:
  R_{in}: $R_{\mathrm{in}} \approx R_{\mathrm{G}} = \mathbf{5\ M\Omega}$
  R_{out}: $R_{\mathrm{out}} \approx R_{\mathrm{D}} \approx \mathbf{1.2\ k\Omega}$
  A_{v}: $A_{\mathrm{V}} = g_{\mathrm{m}}\,(R_{\mathrm{D}} \parallel R_{\mathrm{L}}) = 3$ ms $(1.2$ kΩ $\parallel 10$ kΩ$) \approx \mathbf{3.21}$

8.2.3   *Common-gate amplifier analysis*
Common-gate (CG) circuit
- Gate – common
- V_{in} – source
- V_{out} – drain

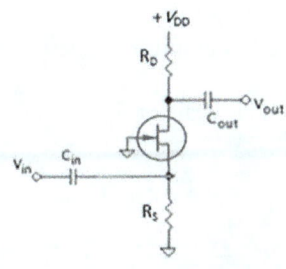

Figure 8.13   CG amplifier

AC equivalent circuit

- AC equivalent circuit of a JFET common-gate (CG) amplifier is obtained with replacing all capacitors by a short circuit and setting DC voltage source V_{DD} to zero and replacing it with a ground.

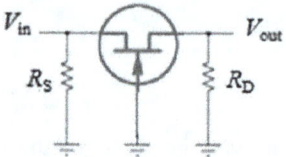

Figure 8.14   AC equivalent circuit of a CG amplifier

Input resistance

- The input resistance at the source of the transistor:

$$R_{is} = \frac{1}{g_m}$$

Derive: $R_{is} \; \dfrac{V_{in}}{I_{in}} = \dfrac{V_{in}}{I_s} \approx \dfrac{V_{in}}{I_d} = \dfrac{V_{gs}}{g_m V_{gs}} = \dfrac{1}{g_m}$ $I_{in} = I_s \approx I_d \approx g_m V_{gs}$, $\left(g_m = \dfrac{I_d}{V_{gs}} \right)$

- The total circuit input resistance presented to the AC source R_{in}:

$$R_{in} = R_S \backslash\backslash R_{is} = R_S \backslash\backslash \frac{1}{g_m}$$

- Unlike CS and CD circuits, the CG configuration provides a low input resistance.

Output resistance

- The output resistance presented to the load of the circuit:
  $R_{out} = R_D \backslash\backslash r_{ds}$
- AC internal resistance between the drain and the source r_{ds} represents reverse-biased drain-source junction resistance. It is typically much larger than R_D and can be replaced by an open circuit ($r_{ds} \approx \infty$).
  $R_{out} \approx R_D$

Voltage gain

- The voltage gain of JFET CG amplifier: the ratio of output to input in the form of voltage.
- Voltage gain without the load resistor R_L:

$$A_V = \frac{V_{out}}{V_{in}} = g_m R_D$$

Derive: $A_V = \dfrac{V_{out}}{V_{in}} = \dfrac{V_d}{V_{gs}} = \dfrac{I_d R_D}{V_{gs}} = \dfrac{g_m V_{gs} R_D}{V_{gs}} = g_m R_D$ $I_d \approx g_m V_{gs}$

- Voltage gain with the load resistor R_L:

$$A_V = \frac{V_{out}}{V_{in}} = g_m \, (R_D \,\|\, R_L)$$

Phase relationship
- In phase: the output voltage in a CG amplifier is in phase with the input voltage.
- The CG amplifier receives its input signal to the source with the output voltage taken from across the drain load ($I_d \approx I_s$, the drain signal "follows" the source signal).

Example: For the JFET CG amplifier of Figure 8.13, determine the input and output resistance R_{in}, R_{out}, and voltage gain A_V, if R_S is 5 kΩ, R_D is 5 kΩ, R_L is 20 kΩ, V_{DD} is 15 V, and g_m is 3 ms.
- Given: $R_S = 5$ kΩ, $R_D = 5$ kΩ, $R_L = 20$ kΩ, $V_{DD} = 15$ V, and $g_m = 3$ ms (Y_{fs} – datasheets).
- Find: R_{in}, R_{out}, and A_V.
- Solution:

$$\mathbf{R_{in}}: R_{in} = R_S \,\|\, \frac{1}{g_m} = 5\,\text{kΩ} \,\|\, \frac{1}{3\text{ms}} \approx \mathbf{0.31\ kΩ} \qquad \text{Kilo: } 10^3;\ \text{Milli: } 10^{-3}$$

$$\mathbf{R_{out}}: R_{out} \approx R_D \approx \mathbf{5\ kΩ}$$

$$\mathbf{A_v}: A_V = g_m \, (R_D \,\|\, R_L) = 3\ \text{ms} \,(5\ \text{kΩ} \,\|\, 20\ \text{kΩ}) \approx \mathbf{12}$$

Summary

FET amplifier
- FET amplifier: an FET in a linear region around an operating point can work as an amplifier to amplify a small AC input signal (low-level signal such as a wireless signal).
- A small-signal JFET (N-channel) amplifier with voltage-divider bias:

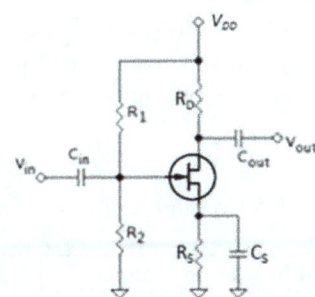

Figure 8.15   A JFET amplifier with voltage-divider bias

Transconductance

- Transconductance (g_m): $g_m = \dfrac{\text{change in } I_{out}}{\text{change in } V_{in}}$ $g_m = y_{fs}$ (datasheets)

- Transconductance for FET:

  AC: $g_m = \dfrac{I_d}{V_{gs}}$

  DC: $g_m = \dfrac{\Delta I_d}{\Delta V_{gs}}$

JFET (N-channel) AC equivalent circuit:

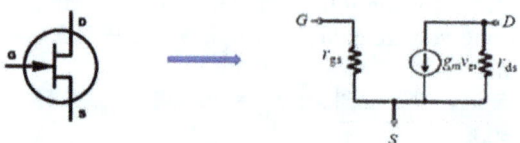

Figure 8.16   JFET AC equivalent circuit

FET internal AC resistance

- Input resistance r_{gs}: the FET internal resistance between the gate and the source. It is high enough to be approximated as an open circuit ($r_{gs} \approx \infty$).
- Output resistance r_{ds}: the FET internal resistance between the drain and the source. it is typically much larger than R_D and can be neglected in calculations (it can be replaced by an open circuit, $r_{ds} \approx \infty$).

Simplified JFET equivalent circuit:

- r_{gs} and r_{ds}: very large – open circuit
- The drain branch acts as a current source ($I_d = g_m v_{gs}$). $g_m = \dfrac{I_d}{V_{gs}}$

The Q-point of JFET

- The Q-point of a JFET circuit:
  Bias a JFET → establish a Q-point (I_D and V_{DS})
- Midpoint biasing: an amplifier with a centered Q-point on the DC load line can provide amplification of an AC input signal without distortion or clipping to the output waveform.

Table 8.1   Three JFET amplifier configurations

Amplifier	Abbreviation	Input	Output	Common terminal
Common-source amplifier	CS	Gate	Drain	Source
Common-drain amplifier	CD	Gate	Source	Drain
Common-gate amplifier	CG	Source	Drain	Gate

Configuration	A_v (without R_L)	A_v (with R_L)	R_{in}	R_{out}	Phase relationship
Common source	$A_V = g_m R_D$	$A_V = g_m (R_D \parallel R_L)$	$R_{in} \approx R_G$	$R_{out} \approx R_D$	180° out of phase
Common drain	$A_V = \dfrac{g_m R_S}{1 + g_m R_S}$	$A_V = \dfrac{g_m (R_S \parallel R_L)}{1 + g_m (R_S \parallel R_L)}$	$R_{in} \approx R_G$	$R_{out} \approx \dfrac{1}{g_m} \parallel R_S$	In phase
Common gate	$A_V = g_m R_D$	$A_V = g_m (R_D \parallel R_L)$	$R_{in} = R_S \parallel \dfrac{1}{g_m}$	$R_{out} \approx R_D$	In phase

Configuration	Circuit	AC equivalent circuit
Common source Input – gate Output – drain		
Common drain Input – gate Output – source		
Common gate Input – source Output – drain		

Self-test

8.1

1. The transconductance for FET is the ratio of the change in drain current to the change in (   ) voltage.
2. The input resistance of the FET is the internal resistance between the gate and the (   ).
3. The output resistance of the FET is the internal resistance between the (   ) and the source.
4. The Q-point of a JFET circuit is obtained from the values of the (   ) current versus the drain-source voltages.

8.2

5. The CD amplifier is also known as the (   ) follower, because the load voltage "follows" the input signal very closely, and voltage gain is approximately equal to 1.

6. The CD amplifier receives its input signal to the gain with the output voltage taken from across the (   ).

7. The value of (   ) resistance on a CD amplifier is usually very high.

8. For the JFET CD amplifier of Figure 8.17, determine the input and output resistance R_{in}, R_{out}, and voltage gain A_V, if R_G is 9 MΩ, R_S is 6 kΩ, R_L is 14 MΩ, V_{DD} is 15 V, and g_m is 3 ms.

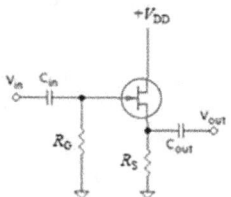

Figure 8.17   Ch 8: No. 8, self-test

9. The output voltage in a CS amplifier is (   ) phase with the input voltage.

10. For the JFET CD amplifier of Figure 8.18, determine the input and output resistance R_{in}, R_{out}, and voltage gain A_V, if R_G is 4 MΩ, R_D is 1 kΩ, R_L is 8 kΩ, and g_m is 4 ms.

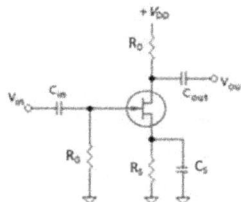

Figure 8.18   Ch 8: No. 10, self-test

11. The output voltage in a common-gate amplifier is ( ) phase with the input voltage.

12. For the JFET CG amplifier of Figure 8.19, determine the input and output resistance R_{in}, R_{out}, and voltage gain A_V, if R_S is 6 kΩ, R_D is 5 kΩ, R_L is 18 kΩ, and g_m is 4 ms.

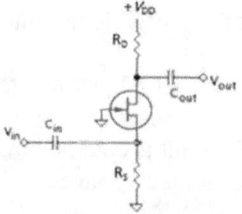

Figure 8.19   Ch 8: No. 12, self-test

Chapter 9

Operational amplifiers

Chapter outline

9.1   Operational amplifiers (op-amps)

9.1.1   Introduction to op-amps

Op-amp

- Recall – amplifier: an electronic device that increases the voltage, current, or power of an input signal.
- Operational amplifier (op-amp): an integrated circuit (IC) that operates as an amplifier to amplify weak electric signals. It is a voltage-controlled amplifier with very high gain.
- IC: an electronic device formed on a small flat piece (chip) of semiconductor material (normally silicon) that can hold many electronic components (hundreds to millions of transistors, resistors, capacitors, etc.).

Symbol and terminals of an op-amp

- Op-amp schematic symbol:
    - An op-amp circuit with two inputs and one output.
    - Inverting input: the terminals labeled with the "–".
    - Non inverting input: the terminals labeled with the "+."

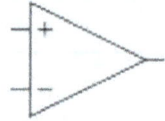

Figure 9.1   Op-amp symbol

- Op-amp schematic symbol with DC power supplies:

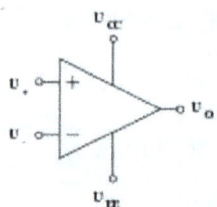

Figure 9.2   Op-amp with DC power supplies

Ideal op-amp vs. practical op-amp

- Ideal (or perfect) op-amp: an op-amp with certain special characteristics to allow it to function with high efficiency.

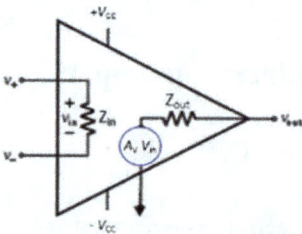

Figure 9.3   Characteristics of an op-amp

- Characteristics of an ideal op-amp:
    - Infinite open-loop gain: $A_V = \infty$
        Open-loop gain: the gain of the op-amp without feedback.
    - Infinite input impedance: $Z_{in} = \infty$
    - Zero output impedance: $Z_{out} = 0$
    - Infinite bandwidth (BW): $BW = \infty$
    - Zero input offset voltage: Exactly zero out if zero in.

- Characteristics of a practical op-amp:
  - High open-loop gain: A_v is high
  - High input impedance: Z_{in} is high   It draws very little current and absorbs less power.
  - Low output impedance: Z_{out} is low   It can provide high output current and reduce loading.
  - Limited BW: BW is limited   It can respond to a wide frequency range.

Table 9.1   Ideal op-amp vs. practical op-amp

Op Amp	A_v	Z_{in}	Z_{out}	BW	Circuit
Idea op-amp	$A_V = \infty$	$Z_{in} = \infty$	$Z_{out} = 0$	$BW = \infty$	
Practical op-amp	A_V is high	Z_{in} is high	Z_{out} is low	BW is limited	

The golden rules of op-amps

For an op-amp with external negative feedback:

- The current rule: no current flows into the two inputs of the op-amp.
  $I_{in} = 0$ $\because Z_{in}$ is high.
  It doesn't load the source and does not affect the input voltage.

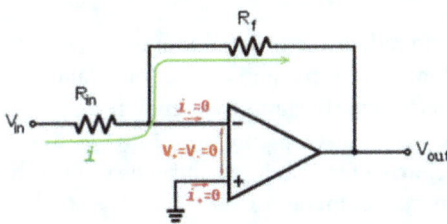

Figure 9.4   The golden rules of op-amps

- The voltage rule: the potential difference between two inputs is zero in an op-amp circuit with negative feedback. $V_{in\,(-)} = V_{in\,(+)}$
  The op-amp will adjust its output to keep the voltage difference between two inputs zero when feedback is being used.

Example:

If the non inverting input $V_{in(+)}$ is connected to ground (set to 0V), i.e. $V_{in(+)} = 0V$, then the inverting input $V_{in(-)}$ is also at 0V, e.g. $V_{in(-)} = 0V$.

Based on the voltage rule: $V_{in(-)} = V_{in(+)}$

$\therefore V_{in(-)} = V_{in(+)} = 0$ V

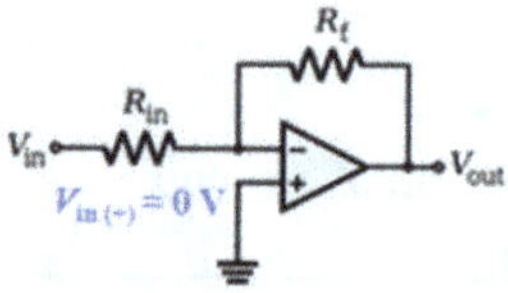

Figure 9.5   The voltage rule – an example

Negative feedback

- Feedback: the process when all or a portion of the output signal is used as an input. It produces a path from the output back to the input.
- There are two types of feedback – positive feedback and negative feedback.
- Positive feedback: when the occurrence of a reaction that enhances or amplifies the original action.
  - The output is connected to the noninverting input $V_{in(+)}$.
  - The fed-back signal is in phase with the input signal.
- Negative feedback: when the occurrence of a reaction reduces the original action and brings the system back to a stable state.
  - The output is connected to the inverting input $V_{in(-)}$.
  - The fed-back signal is out of phase with the input signal.

Benefits of negative feedback

- Promote stability.
- Limit saturation and reduce distortion and noise.
- Increase BW and improve input and output impedances.
- Improve performance (equilibrium, frequency response, step response, etc.).
  Because of these advantages, many op-amps use negative feedback.
  An example of negative feedback: a central heating system allows the room to rest at a predetermined temperature.
- As the temperature rises, the thermostat (the negative feedback part) turns off the furnace.
- As the temperature falls, the thermostat turns on the furnace.

Negative feedback in op-amp

- If op-amp output voltage V_{out} decreases, then inverting input voltage $V_{in(-)}$ will fall by the same amount so differences in the inputs ($V_{in(+)} - V_{in(-)}$) will increase.

$V_{out} \downarrow = V_{in(-)} \downarrow \quad \rightarrow \quad (V_{in(+)} - V_{in(-)}) \uparrow \qquad V_{in(-)} = V_{in(+)}$

- This causes the output voltage V_{out} to rise:

$$V_{out} \uparrow = A_V (V_{in(+)} - V_{in(-)}) \uparrow \qquad A_V = \frac{V_{out}}{V_{in}}$$

- Negative feedback in op-amp:

$$V_{out} \downarrow \rightarrow V_{in(-)} \downarrow \rightarrow (V_{in(+)} - V_{in(-)}) \uparrow$$
$$V_{out} \uparrow$$

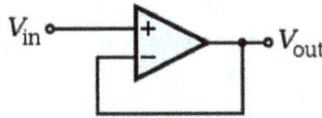

Figure 9.6   Negative feedback in op-amp

9.1.2   Op-amp – modes of operation

Op-amp – three modes of operation

The op-amp exhibits three modes of operation based on the kind of input signals. The three input modes are:

- Single-ended input mode
- Double-ended (differential) input mode
- Common input mode

Single-ended input operation

- Single-ended input mode: input signal is applied only to one of the inputs (one of the inputs is grounded).
- If the input is applied to the non inverting input, the input and output signals are in phase.
- If the input is applied to inverting input, the input and output signals are out of phase.

Figure 9.7   Single-ended input mode

Double-ended input operation

- Double-ended (differential) input mode: apply two opposite polarity signals to two inputs (two inputs are out of phase).

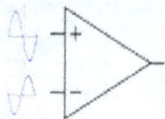

Figure 9.8   Double-ended input mode

- Double-ended input mode amplifies the difference between both signals. The resultant output signal is a superimposing of two input signals.
  It has a peak value twice the value for a single input signal.

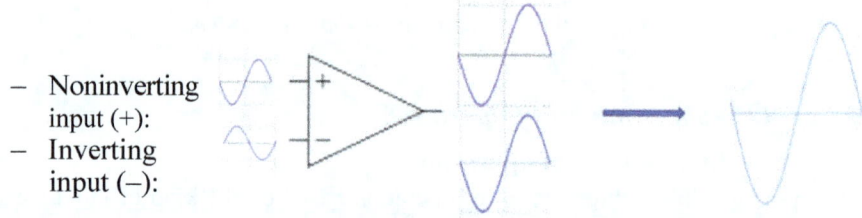

- Noninverting input (+):
- Inverting input (−):

Figure 9.9   Superimposing of two input signals

Common-mode input operation

- Common-mode input: two identical signals appear on both inputs (two inputs are identical in both magnitude and phase).
- In an ideal op-amp, the common-mode signal (unwanted signal) will be canceled out.

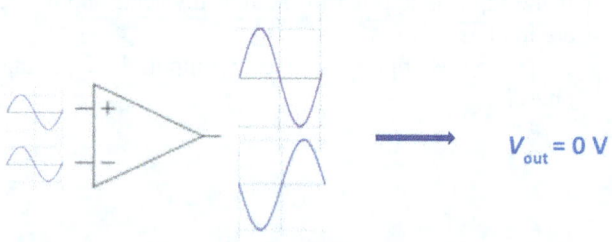

$$V_{out} = 0\ V$$

Figure 9.10   Common-mode input

Table 9.2:   *The input modes of op-amp*

Single-ended input	Double-ended input (differential input)	Common-mode input
- The input signal is applied to one of the inputs. - If the input is applied to the inverting input, the input and output signals are out of phase. - If input is applied to the noninverting input, the input and output signals are in phase.	- Apply two opposite-polarity signals to two inputs. - The output signal is a superimposing of two input signals.	- Two identical signals (noise) appear on both inputs. - In an ideal op-amp, the common-mode signal will be canceled out.

9.2   Basic building blocks of an op-amp

9.2.1   *Inverting op-amp*

Introduction to inverting op-amp

- Inverting op-amp: an op-amp circuit which inverting input (−) receives the input signal and feedback from the output of the op-amp. It has 180° out-of-phase output with respect to the input.
- Inverting op-amp circuit: Input → (−)

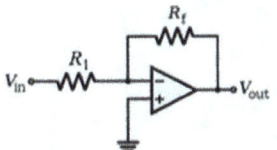

Figure 9.11   *Inverting op-amp*

The closed-loop voltage gain of an inverting op-amp

$$A_{CL} = -\frac{R_f}{R_1}$$
    The voltage gain is set by the ratio of two resistors.

Derive: $A_{CL} = \dfrac{V_{out}}{V_{in}} = -\dfrac{I_f R_f}{I_{in} R_1} = -\dfrac{R_f}{R_1}$

$V_{out} = -I_f R_f$ since $V_{in(-)} = 0$, the point A is a virtual ground.

The noninverting input is connected to ground.

$$V_{in(-)} = V_{in(+)} = 0$$

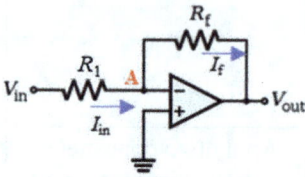

Figure 9.12   Derive the voltage gain

The impedance of an inverting op-amp

- Input impedance of an inverting op-amp Z_i:

$$Z_i = R_1$$

$R_1 = \dfrac{V_{in}}{I_{in}}$, since R_1 is connected to virtual ground (point A) in the inverting input terminal.

- The output impedance of an inverting op-amp Z_o (with feedback):

$$Z_o = \frac{Z_{out}}{1 + A_{OL}\dfrac{R_1}{R_1 + R_f}}$$

Z_{out} – the internal output impedance (without feedback)

A_{OL} – the open-loop voltage gain (without feedback)

Derive: $Z_o = \dfrac{V_{out}}{I_{out}}$

$V_{out} = A_{ol} V_{diff} - Z_{out} I_{out}$

If $A_{ol} V_{diff} \gg Z_{out} I_{out}$

$V_{out} \approx A_{ol} V_{diff} = A_{ol}(V_{in} - V_f)$

$\qquad = A_{ol}\left(V_{in} - V_{out}\dfrac{R_1}{R_1 + R_f}\right)$

$\qquad = A_{ol} V_{in} - A_{ol} V_{out}\dfrac{R_1}{R_1 + R_f}$

$A_{ol} V_{in} = V_{out} + A_{ol} V_{out}\dfrac{R_1}{R_1 + R_f}$

$\qquad = V_{out}\left(1 + A_{ol}\dfrac{R_1}{R_1 + R_f}\right)$

$\qquad = I_{out} Z_o\left(1 + A_{ol}\dfrac{R_1}{R_1 + R_f}\right)$

$A_{CL} = \dfrac{V_{out}}{I_{out}}$, KVL

$V_{diff} = V_{in} - V_f$

$V_f = V_{out}\dfrac{R_1}{R_1 + R_f}$

$$\frac{A_{ol}\,V_{in}}{I_{out}} = \frac{I_{out}\,Z_o}{I_{out}}(1 + A_{ol}\frac{R_1}{R_1 + R_f}) \qquad \div\, I_{out}\ \text{both sides}$$

$$\frac{A_{ol}\,V_{in}}{I_{out}} = Z_o(1 + A_{ol}\frac{R_1}{R_1 + R_f})$$

$$\frac{V_{out}}{I_{out}} = Z_o(1 + A_{ol}\frac{R_1}{R_1 + R_f}) \qquad V_{out} \approx A_{ol}V_{in}\ \text{(if without feedback)}$$

$$Z_{out} = Z_o(1 + A_{ol}\frac{R_1}{R_1 + R_f}) \qquad Z_{out} = \frac{V_{out}}{I_{out}}$$

$$\therefore Z_o = \frac{Z_{out}}{1 + A_{OL}\frac{R_1}{R_1 + R_f}}$$

Z_o (with feedback) is less than Z_{out} (without feedback). $\qquad Z_o\downarrow = \dfrac{Z_{out}}{(1 + A_{OL}\frac{R_1}{R_1 + R_f})\uparrow}$

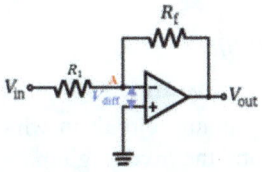

Figure 9.13   Derive the output impedance

Characteristics of an inverting op-amp

- The closed-loop voltage gain of an inverting op-amp is negative ($A_{CL} = -\dfrac{R_f}{R_1}$), indicating that the output is out of phase with the applied input signal.
- An inverting op-amp provides stability to the system since it is with negative feedback (the output is connected to the inverting input $V_{in\,(-)}$).
- The voltage gain of an inverting op-amp (A_{CL}) is set by resistor ratio ($-\dfrac{R_f}{R_1}$). This indicates that it is hard to achieve high voltage gain with this circuit.

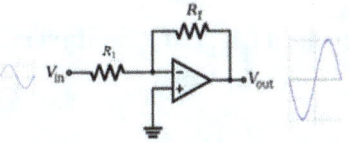

Figure 9.14   Phase relation

Example: Determine the output voltage, and input and output impedance for the circuit shown in Figure 9.11, if V_{in} is 50 mV, R_1 is 2kΩ, R_f is 50kΩ, Z_{out} is 80Ω, and A_{OL} is 150,000.

- Given: $V_{in}= 50$ mV, $R_1= 2$ kΩ, $R_f= 50$ kΩ, $Z_{out}= 80Ω$, and $A_{OL} = 150{,}000$.
- Find: V_{out}, Z_i, and Z_o.
- Solution:

$$V_{out}: V_{out} = A_{CL}V_{in} = -\frac{R_f}{R_1}V_{in}$$

$$A_{CL} = \frac{V_{out}}{V_{in}}, \quad A_{CL} = -\frac{R_f}{R_1}$$

$$= (-\frac{50\ kΩ}{kΩ})(50\ mV) \approx -1{,}250\ mV$$

$$Z_i: Z_i = R_1 = 2\ kΩ$$

$$Z_o: Z_o = \frac{Z_{out}}{1+A_{OL}\dfrac{R_1}{R_1+R}} = \frac{80\ Ω}{1+150{,}000\dfrac{2Ω}{2Ω+50Ω}} \approx 0.01386Ω = 13.86mΩ$$

milli: 10^{-3}

9.2.2   Noninverting op-amp

Introduction to noninverting op-amp

- Noninverting op-amp: an op-amp circuit in which the noninverting input (+) receives the input signal, and the inverting input (−) receives positive feedback from the output of the op-amp. The output signal is in phase with the input signal applied.
- Noninverting op-amp circuit:

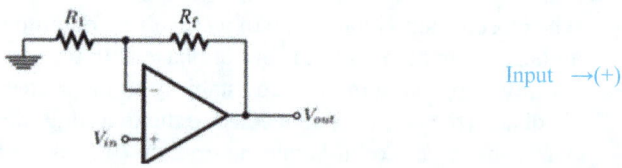

Input →(+)

Figure 9.15   *Noninverting op-amp*

The closed-loop voltage gain (A_{CL}) of a noninverting op-amp:

- $$A_{CL} = 1 + \frac{R_f}{R_1}$$ The voltage gain is set by resistor ratio.

Derive: $V_{in(-)} = V_{in(+)}$ The golden rule of op-amp (voltage rule)

$$V_{in(-)} = V_{out}\frac{R_1}{R_1 + R_1} = V_{in(+)}$$ Voltage divide rule

$$A_{CL} = \frac{V_{out}}{V_{in}} = \frac{R_1 + R_f}{R_1} = 1 + \frac{R_f}{R_1}$$

The impedance of a noninverting op-amp

- The input impedance of a noninverting op-amp Z_i:

$$Z_i = Z_{in}\left(A_{OL}\frac{R_1}{R_1 + R_f} + 1\right)$$

Z_i – with feedback, Z_{in} –without feedback

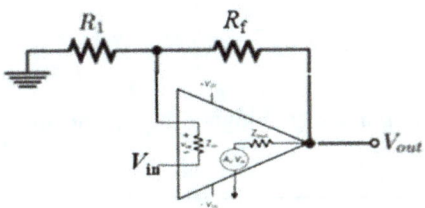

Figure 9.16   Input impedance of a noninverting op-amp

Derive: $V_{in} = V_f + V_{diff}$

$$= \frac{R_1}{R_1 + R_f}V_{out} + V_{diff}$$

$$= \frac{R_1}{R_1 + R_f}A_{OL}V_{diff} + V_{diff}$$

$$= V_{diff}\left(\frac{R_1}{R_1 + R_f}A_{OL} + 1\right)$$

$$V_{in} = I_{in}Z_{in}\left(\frac{R_1}{R_1 + R_f}A_{OL} + 1\right)$$

$$\frac{V_{in}}{I_{in}} = Z_{in}\left(\frac{R_1}{R_1 + R_f}A_{OL} + 1\right)$$

$$\therefore Z_i = \frac{V_{in}}{I_{in}} = Z_{in}\left(\frac{R_1}{R_1 + R_f}A_{OL} + 1\right)$$

$$V_f = V_{out}\frac{R_1}{R_1 + R_f}$$

$$A_{OL} = \frac{V_{out}}{V_{diff}}, \quad V_{out} = A_{OL}V_{diff}$$

$$V_{diff} = I_{in}Z_{in}$$

$\div I_{in}$ both sides

- The output impedance of a noninverting op-amp Z_o:

$$Z_o = \frac{Z_{out}}{1 + A_{OL}\dfrac{R_1}{R_1 + R_f}}$$

Z_o – with feedback, Z_{out} – without feedback

It is the same as the output impedance of an inverting op-amp.

Characteristics of a noninverting op-amp:

- Closed-loop voltage gain is positive, indicating that the output is in phase with the input signal.

$$A_{CL} = 1 + \frac{R_f}{R_1}$$

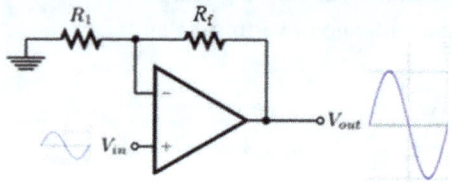

Figure 9.17   Phase relation

- The noninverting amp has a gain higher than 1, meaning that the output signal is amplified.

Example: Determine the output voltage, input and output impedance for the circuit shown in Figure 9.15, if V_{in} is 50 mV, R_1 is 3 kΩ, R_f is 50 kΩ, A_{OL} is 150,000, Z_{in} is 1.5 MΩ, and Z_{out} is 50 Ω.

- Given: $V_{in} = 50$ mV, $R_1 = 3$ kΩ, $R_f = 50$ kΩ, $A_{OL} = 150,000$, $Z_{in} = 1.5$ MΩ and $Z_{out} = 50$ Ω.
- Find: V_{out}, Z_i, and Z_o
- Solution:

V_{out}: $V_{out} = A_{CL} V_{in} = (1 + \dfrac{R_f}{R_1} V_{in})$ $\qquad A_{CL} = \dfrac{V_{out}}{V_{in}}, A_{CL} = 1 + \dfrac{R_f}{R_1}$

$= (1 + \dfrac{50\ \text{k}\Omega}{3\ \text{k}\Omega}) (50\ \text{mV}) \approx \mathbf{883.3\ mV}$

Z_i: $Z_i = (A_{OL} \dfrac{R_1}{R_1 + R_f} + 1) Z_{in}$

$= (150,000 \dfrac{3\ \text{k}\Omega}{3\ \text{k}\Omega + 50\ \text{k}\Omega} + 1)\ 1.5\ \text{M}\Omega$

$\approx 12737.35\ \text{M}\Omega \approx \mathbf{12.74\ G\Omega}$ Mega: 10^6:Giga: 10^9

$Z_o = \dfrac{Z_{out}}{1 + A_{OL} \dfrac{R_1}{R_1 + R_f}} = \dfrac{50\ \Omega}{1 + 150,000 \dfrac{3\text{k}\Omega}{3\ \text{k}\Omega + 50\ \text{k}\Omega}}$

$\approx 0.005888\ \Omega \approx \mathbf{5.9\ m\Omega}$ Milli: 10^{-3}

Op-amp: high input impedance and low output impedance.

9.2.3   Op-amp

Introduction to voltage follower op-amp

- Voltage follower op-amp (buffer, unity gain amplifier, isolation amplifier): an op-amp circuit configuration that has a gain of 1.

- Voltage follower op-amp circuit:

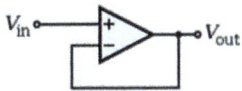

Figure 9.18   Voltage follower

The closed-loop voltage gain of a voltage follower:

$A_{CL} = 1$

Derive: $V_{in(-)} = V_{in(+)} = V_{out}$ Golden rules of op-amp (voltage rule)

$A_{CL} = \dfrac{V_{out}}{V_{in}} = 1$ $V_{in(-)} = V_{in(+)} = V_{in}$

The impedance of a voltage follower

- The input impedance of a voltage follower Z_i:

$Z_i = Z_{in}(A_{OL} + 1)$

Derive: recall the input impedance of a noninverting op-amp is

$Z_i = Z_{in}(A_{OL} \dfrac{R_1}{R_1 + R_f} + 1)$

A voltage follower is a special case of the noninverting op-amp with $R_f = 0$ (short circuit), and R_1 is removed (open circuit).

$Z_i = Z_{in}(A_{OL} \dfrac{R_1}{R_1 + 0} + 1)$ $R_f = 0$

$\therefore Z_i = Z_{in}(A_{OL} + 1)$

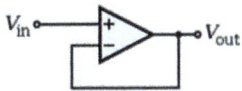

Figure 9.19   Derive the voltage follower

- The output impedance of a voltage follower Z_o:

$Z_o = A_{OL} \dfrac{Z_{out}}{1 + A_{OL}}$

Derive: recall the input impedance of a noninverting op-amp is

$Z_o = \dfrac{Z_{out}}{1 + A_{OL}\dfrac{R_1}{R_1 + R_f}}$

$$Z_o = \frac{Z_{out}}{1 + A_{OL} \dfrac{R_1}{R_1 + 0}} \qquad\qquad R_f = 0$$

$$Z_o = \frac{Z_{out}}{1 + A_{OL}}$$

Characteristics of a voltage follower

- The output of the voltage follower "follows" the input signal (output voltage is equal to the input voltage). $V_{in} = V_{out}$ $\qquad A_{CL} = \dfrac{V_{out}}{V_{in}} = 1$

- A voltage follower is a special case of the noninverting amplifier with the feedback resistor $R_f = 0$ (a short circuit) and input resistor $R_{in} = \infty$ (open circuit).
- The high input and low output impedances of the voltage follower make it a good "buffer" between two circuits.
  - The voltage follower has a very high input impedance that will ensure maximum current is provided by the source (the op-amp takes less current from the input).
  - The voltage follower has a very low output impedance that will make all the voltage available at the output (it does not cause a loading effect).

Example: Determine the output voltage, and input and output impedance for the circuit shown in Figure 9.18, if V_{in} is 50 mV, A_{OL} is 150,000, Z_{in} is 1.5 MΩ and Z_{out} is 50 Ω.

- Given: $V_{in} = 50$ mV, $A_{OL} = 150,000$, $Z_{in} = 1.5$ MΩ, and $Z_{out} = 50$ Ω.
- Find: V_{out}, Z_i, and Z_o.
- Solution:

  V_{out}: $V_{out} = A_{CL} V_{in} = (1)(V_{in}) = (1)\,(50\text{ mV}) = \mathbf{50\ mV}$ $\qquad A_{CL} = 1$

  $\mathbf{Z_i}$: $Z_i = Z_{in}(A_{OL} + 1) = 1.5$ MΩ $(150,000 + 1) \approx 225,002$ M$\Omega \approx \mathbf{225\ G\Omega}$

  $\mathbf{Z_i}$: $Z_o = \dfrac{Z_{out}}{1 + A_{OL}} = \dfrac{50\ \Omega}{1 + 150,000} \approx 0.000333\Omega = \mathbf{333\ \mu\Omega}$ $\qquad$ Micro(μ): 10^{-6}

Table 9.3   Basic op-amp circuits

	Noninverting op-amp	Inverting op-amp	Voltage follower
Circuit	Input → (+)	Input → (−)	
Formulas	$A_{CL} = 1 + \dfrac{R_f}{R_1}$ $Z_i = Z_{in}(A_{OL}\dfrac{R_1}{R_1 + R_f} + 1)$ $Z_o = \dfrac{Z_{out}}{1 + A_{OL}\dfrac{R_1}{R_1 + R_f}}$	$A_{CL} = -\dfrac{R_f}{R_1}$ $Z_i = R_1$ $Z_o = \dfrac{Z_{out}}{1 + A_{OL}\dfrac{R_1}{R_1 + R_f}}$	$A_{CL} = 1$ $Z_i = Z_{in}(A_{OL} + 1)$ $Z_o = \dfrac{Z_{out}}{1 + A_{OL}}$
Phase relation-ship	The input and output signals are in phase.	The input and output signals are out of phase.	The output "follows" the input signal.

9.2.4   Terminologies of op-amps

Glossary of op-amp key terms

- Differential input voltage (V_{diff}):the maximum voltage that can be applied to two input terminals of an op-amp without damaging it.
- Open-loop voltage gain (A_{OL}):the voltage gain of the op-amp when no feedback is applied to the circuit (open-loop format).

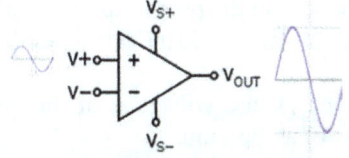

Figure 9.20   Open-loop voltage gain

A_{OL} is determined by the internal component values and datasheet values.

- Closed-loop voltage gain (A_{CL}): the voltage gain of the op-amp when negative feedback is applied to the circuit (closed-loop format).

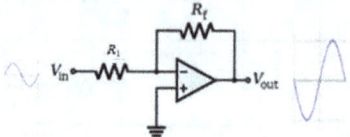

Figure 9.21   Closed-loop voltage gain

A_{CL} is determined by the external component values and can be controlled by external components.

- Input impedance (Z_{in}): the impedance viewed (seen) from the input terminal of the op-amp.
  - Common-mode input impedance: the input impedance of each input of an op-amp with respect to ground.
  - Differential input impedance: the total impedance between two inputs of an op-amp.

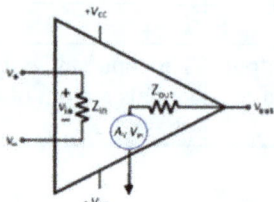

Figure 9.22   Differential input impedance

- Output impedance (Z_{out}): the impedance viewed from the output terminal of an op-amp.
- Peak-to-peak output voltage swing: the maximum peak-to-peak output voltage that an op-amp physically provides at its output without waveform clipping.
- Common-mode voltage range (CMVR) or input voltage range (IVR): a range of input voltages that produces in the proper operation of an op-amp.
- Common-mode signals (noise): an identical signal that appears on both inputs of an op-amp.
- Common-mode gain (A_{CM}): the voltage gain for the common-mode signals between the two inputs of an op-amp.
- Common-mode rejection: the ability of an op-amp to eliminate common-mode signals to both inputs.
  An unwanted signal that appears on inputs will not appear on the outputs.
- Common-mode rejection ratio (CMRR): the measure of the ability of an op-amp to eliminate common-mode signals (noise). It is the ratio of the open-loop gain (or differential gain) to the common-mode gain.   Higher CMRR is better.

- Calculate CMRR: $\text{CMRR} = \dfrac{\text{open-loop gain}}{\text{common-mode gain}} = \dfrac{A_{OL}}{A_{CM}}$

- CMRR in dB: $\qquad \text{CMRR} = 20 \log \left(\dfrac{A_{OL}}{A_{CM}}\right)$

- Slew rate: the maximum rate of change in the output of an op-amp caused by a step change (such as 1V) on the input(per unit of time).
- Input offset voltage (V_{os}): the DC voltage applied to the input of an op-amp that makes the output voltage exactly zero.
  - Ideal op-amp: $\qquad V_{in} = 0 \rightarrow V_{out} = 0$
  - Practical op-amp: $V_{in} = 0 \rightarrow V_{out} \neq 0$ (it drifts with temperature, $T \rightarrow V_{os}$)

  When the inputs are grounded, ideally the output of the op-amp should be at zero, but practically it is not zero. To make this output voltage zero, a small amount of input offset voltage is required.

  Many op-amps have pins available for an offset null that can effectively apply V_{os} to ensure that the offset is removed from the output.

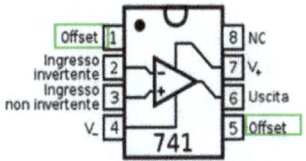

Figure 9.23   Offset null

- Input offset current (I_{os}): a difference in the input current that flows in or out of each of the input pins of an op-amp even if the output voltage is exactly zero.
  - Ideal op-amp: $\qquad I_{in(+)} = I_{in(-)} = 0 \rightarrow I_{in(+)} - I_{in(-)} = 0$
  - Practical op-amp: $I_{in(+)} \neq I_{in(-)} \neq 0 \rightarrow I_{in(+)} - I_{in(-)} \neq 0$

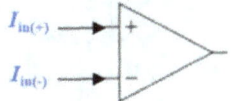

Figure 9.24   Input currents

- Input bias current (I_{inB}): the DC current required by two input terminals of an op-amp with the output at a specified level (to properly operate the first stage). It is defined as the average of the two input currents.

$$I_{inB} = \dfrac{I_{in(+)} + I_{in(-)}}{2}$$

- Bias current compensation in a voltage follower: add a resistor (R_{comp}) equal to the source resistor (R_s) in the feedback path ($R_{comp} = R_s$).
- Bias current compensation in an inverting or noninverting op-amp: add a resistor (R_{comp}) equal to the input resistor (R_{in}) in parallel with the feedback resistor (R_f) in the input path ($R_{comp} = R_{in} \setminus\setminus R_f$).

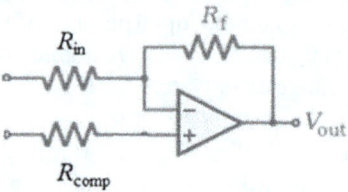

Figure 9.25   Bias current compensation

- Op-amp frequency response: the change in gain of an op-amp to respond to a change in the frequency of the input signal.

9.3   Basic op-amp circuits

9.3.1   *Summing amplifier*

Introduction to summing amplifier

- Summing amplifier (adder): an op-amp circuit configuration which can be used to sum signals (as the name suggests). It can combine two or more input signals to a single output.
- Summing amplifier circuit:

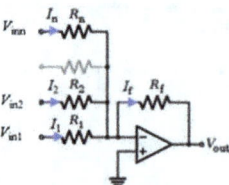

Figure 9.26   Summing amplifier

The output voltage of a summing amplifier

- Output voltage: $V_{out} = -(\dfrac{V_{in1}}{R_1} + \dfrac{V_{in2}}{R_2} + \dfrac{V_{in3}}{R_3} \dots) R_f$

Derive: $I_f = I_1 + I_2 + I_3 + \dots$ $\qquad\qquad\qquad \sum I = 0$

$V_{out} = -I_f R_f = -(I_1 + I_2 + I_3 + \dots) R_f$

$\qquad = -(\dfrac{V_{in1}}{R_1} + \dfrac{V_{in2}}{R_2} + \dfrac{V_{in3}}{R_3} \dots) R_f$ $\qquad\qquad I = \dfrac{V}{R}$

Characteristics of a summing amplifier

- The summing amplifier uses an inverting amplifier configuration, and the non inverting input terminal is connected to ground.   Input $\rightarrow$ (–)
- Closed-loop voltage gain is negative, indicating that the output will be out of phase with the applied input signal.
- If the resistors, R_1, R_2, R_3, R_f, etc. are all equal, a summing amplifier will have a unity gain.

  If $R_1 = R_2 = R_3 = \ldots = R_f$: $V_{out} = -(V_{in1} + V_{in2} + V_{in3} + \ldots)$

  $V_{out} = -(\dfrac{V_{in1}}{R_1} + \dfrac{V_{in2}}{R_2} + \ldots)\, R_f$

- If the resistors R_1, R_2, R_3, etc. are of different values, a summing amplifier will output a weighted ($\dfrac{R_f}{R_1}$, $\dfrac{R_f}{R_2}$, etc.) sum of the input signals (a scaling adder).

Example: Determine the output voltage for the circuit shown in Figure 9.27, if R_1 is 3 kΩ, R_2 is 1.5 kΩ R_3 is 2 kΩ, R_f is 20 kΩ, V_{in1} is 3 mV, V_{in2} is 2 mV, and V_{in3} is 4 mV.

- Given: $R_1 = 3$ kΩ, $R_2 = 1.5$ kΩ, $R_3 = 2$ kΩ, $R_f = 20$ kΩ, $V_{in1} = 3$ mV, $V_{in2} = 2$ mV, and $V_{in3} = 4$ mV.

- Find: V_{out}

- Solution: $V_{out} = -(\dfrac{V_{in1}}{R_1} + \dfrac{V_{in2}}{R_2} + \dfrac{V_{in3}}{R_3})\, R_f$

  $= -(\dfrac{3\ mV}{3\ kΩ} + \dfrac{4\ mV}{1.5\ kΩ} + \dfrac{4\ mV}{2\ kΩ})\, 20\ kΩ \approx -\mathbf{86.7\ mV}$

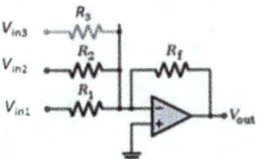

Figure 9.27   *Summing amplifier – an example*

9.3.2   *Difference amplifier*

Introduction to difference amplifier

- Difference amplifier (subtractor): an op-amp circuit configuration which amplifies the difference between two input signals but reject signals that are common to both inputs.

- Difference amplifier circuit:

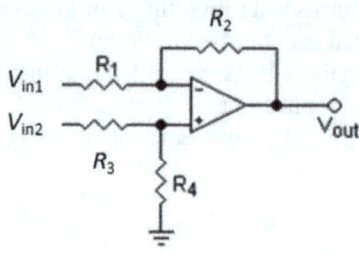

Figure 9.28   Difference amplifier

Characteristics of a difference amplifier

- The difference amplifier is a combination of both inverting and noninverting amplifiers. It has two inputs and one output.   Input $\rightarrow$ (−) and (+)
- Difference amplifiers are used mainly to reject noise. Any signal common to both inputs (common-mode noise) will be automatically canceled out.
  Common mode signals (noise): an identical signal that appears on both inputs of an op-amp.
- Difference amplifiers have high differential voltage gain (open-loop gain) and very low common mode gain–CMRR is high.

$$\text{CMRR} \uparrow\uparrow = \frac{\text{open-loop gain } \uparrow}{\text{common-mode gain } \downarrow}$$

The output voltage of a difference amplifier

- If $R_1 = R_3$ and $R_2 = R_4$:

  The output voltage: $V_{out} = (V_{in2} - V_{in1}) \dfrac{R_2}{R_1}$

Derive: If $V_{in2} = 0$, $V_{out1} = -V_{in1} \dfrac{R_2}{R_1}$ Inverting op-amp: $A_{CL} = \dfrac{V_{out}}{V_{in}} = -\dfrac{R_f}{R_1}$

− If $V_{in1} = 0$, $V_{R4} = V_{in(-)} = V_{out2} \dfrac{R_1}{R_1 + R_2}$ $V_{in(-)} = V_{in(+)}$

$V_{out2} = V_{R4} \dfrac{R_1 + R_2}{R_1}$ Solve for V_{out2}

$V_{R4} = V_{in2} \dfrac{R_4}{R_3 + R_4}$

$V_{out2} = (V_{in2} \dfrac{R_4}{R_3 + R_4})(\dfrac{R_1 + R_2}{R_1})$ Substitute $V_{R4} = V_{in2} \dfrac{R_4}{R_3 + R_4}$

–   If $R_1 = R_3$, $R_2 = R_4$:

$$V_{out2} = (V_{in2} \frac{R_4}{R_3 + R_4})(\frac{R_3 + R_4}{R_1})$$ Substitute R_1 for R_3 and R_2 for R_4

$$= V_{in2} \frac{R_4}{R_1} = V_{in2} \frac{R_2}{R_1}$$ $R_2 = R_4$

–   If $V_{in2} \neq 0$, $V_{out} = V_{out2} + V_{out1} = V_{in2} \frac{R_2}{R_1} + (-V_{in1} \frac{R_2}{R_1})$ $V_{out1} = -V_{in1} \frac{R_2}{R_1}$

$$\therefore V_{out} = (V_{in2} - V_{in1}) \frac{R_2}{R_1}$$ Factor out $\frac{R_2}{R_1}$

If $R_1 = R_2$:

- The output voltage: $V_{out} = V_{in2} - V_{in1}$ $V_{out} = (V_{in2} - V_{in1}) \frac{R_2}{R_1}$

Example: Determine the output voltage for the difference amplifier circuit shown in Figure 9.28, if R_1 is 1.5 kΩ, R_2 is 2 kΩ, V_{in1} is 2 mV, and V_{in2} is 4 mV.
- Given: $R_1 = 1.5$ kΩ, $R_2 = 2$ kΩ, $V_{in1} = 2$ mV, and $V_{in2} = 4$mV.
- Find: V_{out}
- Solution: $V_{out} = (V_{in2} - V_{in1}) \dfrac{R_2}{R_1} = (4 \text{ mV} - 2\text{mV}) \dfrac{2 \text{ kΩ}}{1.5 \text{ kΩ}} \approx \textbf{2.7mV}$

9.3.3   *Op-amp differentiator*

Introduction to op-amp differentiator

- Op-amp differentiator: an op-amp circuit configuration whose output voltage is proportional to differentiation of input voltage (output is directly proportional to the input voltage's rate of change with respect to time).
  The op-amp differentiator simulates the mathematical operation of differentiation.
- Op-amp differentiator circuit:

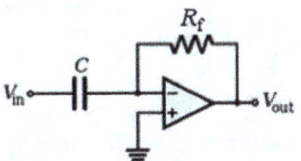

Figure 9.29   *Op-amp differentiator*

Characteristics of a differentiator

- An op-amp differentiator is basically an inverting amplifier in which a resistor provides negative feedback and a capacitor at its input side.
  By replacing the input resistor R_i in the inverting amplifier with a capacitor C, we obtain a differentiator.

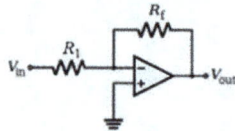

Figure 9.30   Inverting op-amp

- An op-amp differentiator produces a constant output voltage for a steadily changing input voltage.
- Differentiators are commonly used to operate on triangular and rectangular input signals. The output waveform will be changed and whose shape is dependant upon the RC time constant of the differentiator.
- If a triangular input signal (or sawtooth wave) is applied to a differentiator, the output will be changed to a square wave. (An op-amp differentiator performs the mathematical operation of differentiation.)

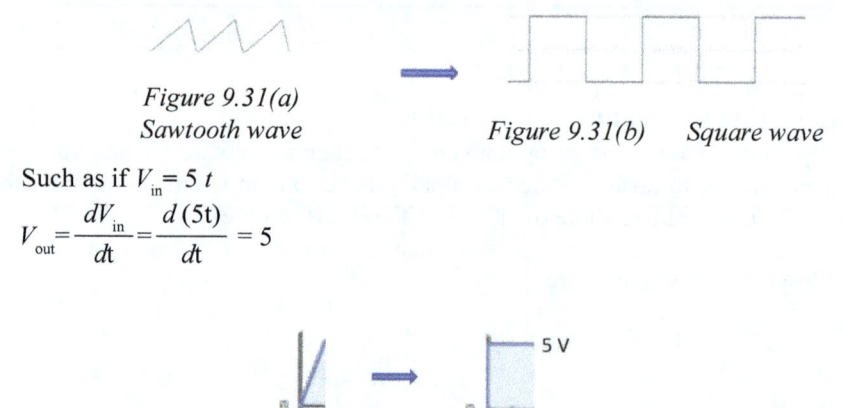

Figure 9.31(a)
Sawtooth wave

Figure 9.31(b)   Square wave

Such as if $V_{in} = 5\,t$

$$V_{out} = \frac{dV_{in}}{dt} = \frac{d\,(5t)}{dt} = 5$$

5 V

Figure 9.32   An example

The output voltage of a differentiating amplifier

$$V_{out} = -R_f C\,\frac{\Delta V_{in}}{\Delta t} \quad \text{or} \quad V_{out} = -R_f C\,\frac{dV_{in}}{dt}$$ $R_f C$ - time constant of the RC circuit

An example of the input voltage's rate of change with respect to time:

$$\frac{\Delta V_{in}}{\Delta t} = \frac{4\ mV - 3\ mV}{3\ ms - 2\ ms} = 1\ mV/ms$$

Derive $V_{out} = R_f C \dfrac{\Delta V_{in}}{\Delta t}$:

$$V_{out} = -I_f R_f = -I_C R_f$$

$$= -(C\frac{\Delta V_C}{\Delta t})R_f = -(C\frac{\Delta V_{in}}{\Delta t})R_f$$

$$\therefore V_{out} = -R_f C \frac{\Delta V_{in}}{\Delta t}$$

$$I_C = I_f$$

$$I_C = C\frac{\Delta V_{in}}{\Delta t},\ V_C = V_{in}$$

The minus sign (−) indicates a 180° phase shift (the input signal is connected to the inverting input).

Example: Determine the output voltage for the differentiator shown in Figure 9.29, if R_f is 3.5 kΩ, C is 0.003 µF, and V_{in} is a triangular input signal rating from −8 V to +8V at 12 µs.

- Given: R_f= 3.5 kΩ, C= 0.003 µF, V_{in} = −8V to +8V, Δt = 12 µs.
- Find: V_{out}
- Solution: ΔV_{in} = 8 V − (−8V) = 16 V

$$V_{out} = -R_f C\frac{\Delta V_{in}}{\Delta t}$$

$$= -(3.5\ k\Omega)(0.003\ \mu F)\frac{16\ V}{12\ \mu s} = -0.014\ KV = -14\ V\ \text{Kilo: }10^3$$

The output voltage is a square wave rating from −14 V to +14V

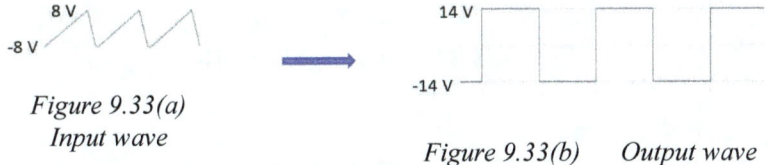

Figure 9.33(a)
Input wave

Figure 9.33(b)   Output wave

9.3.4   Op-amp integrator

Introduction to op-amp integrator

- Op-amp integrator: an op-amp circuit configuration whose output voltage is proportional to the integral of input voltage. (Output is directly proportional to the input voltage integrated over time.)

The op-amp integrator simulates the mathematical operation of integration.

- Op-amp integrator circuit:

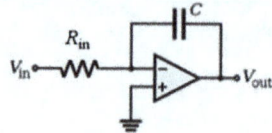

Figure 9.34   Op-amp integrator

Characteristics of an op-amp integrator

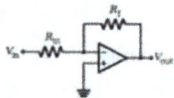

Figure 9.35   Inverting amplifier

- An op-amp integrator is basically an inverting amplifier in which a capacitor provides negative feedback and a resistor at its input side.
  By replacing the feedback resistor R_f in the inverting amplifier with a capacitor C, we obtain an integrator.
- An op-amp integrator produces an output that is proportional to the total area (amplitude times time) included under the waveform.
- If a square wave input signal is applied to an integrator, the output will be changed to a triangular wave.

Figure 9.36   A triangular wave

Figure 9.37(a)   Square wave   *Figure 9.37(b)   Sawtooth wave*

Rate of change of the output voltage of an integrator

$$\frac{\Delta V_{out}}{\Delta t} = -\frac{V_{in}}{R_{in}C}$$

Derive: $I_{in} = I_C$

$$I_{in} = \frac{V_{in}}{R_{in}}$$

$$I_C = - C \frac{\Delta V_C}{\Delta t} = - C \frac{\Delta V_{out}}{\Delta t} \qquad\qquad V_{out} = V_C$$

$$\text{or} \quad I_C = - C \frac{dV_{out}}{dt}$$

$$\frac{V_{in}}{R_{in}} = - C \frac{\Delta V_{out}}{\Delta t} \qquad\qquad \frac{V_{in}}{R_{in}} = I_{in} = I_C = - C \frac{\Delta V_{out}}{\Delta t}$$

$$\therefore \quad \frac{\Delta V_{out}}{\Delta t} = - \frac{V_{in}}{R_{in} C}$$

Example: Determine the rate of change of the output voltage and the waveform for the integrator shown in Figure 9.34, if R_{in} is 8 kΩ, C is 0.03 μF, and V_{in} is a square wave input signal rating from -5V to $+5$V at 500 μs (the wave width). Assume that the voltage across the capacitor is initially zero.

- Given: $R_{in} = 8$ kΩ, $C = 0.03$ μF, $V_{in} = -5$ V to $+5$V, $\Delta t = 500$ μs.
- Find: ΔV_{out}
- Solution:

The rate of change of the output voltage:

- When $V_{in} = + 5$V: $\dfrac{\Delta V_{out}}{\Delta t} = - \dfrac{V_{in}}{R_{in} C} = - \dfrac{5\text{ V}}{(8\text{ k}\Omega)(0.03\text{ μF})}$

$$= - \frac{5\text{ V}}{(8000\ \Omega)(0.00000003\text{ F})}$$
$$\approx -20{,}833\text{ V/s} = -20.833\text{ kV/s} \qquad\qquad \text{Kilo: } 10^3$$
$$\approx -21\text{ mV/ μs} \qquad\qquad \text{Milli: } 10^{-3}; \text{ Micro: } 10^{-6}$$

For $\Delta t = 500$ μs, $\Delta V_{out} \approx (\Delta t)\,(-21\text{ mV/μs})$ $\dfrac{\Delta V_{out}}{\Delta t} \approx -21\text{ mV/μs}$

$$= (500\text{ μs})\,(-21\text{ mV/μs}) = -10{,}500\text{ mV} = \mathbf{-10.5V}$$

- When $V_{in} = -5$V: $\dfrac{\Delta V_{out}}{\Delta t} = - \dfrac{V_{in}}{R_{in} C} = - \dfrac{-5\text{ V}}{(8\text{ k}\Omega)(0.03\text{ μF})} =$

$$- \frac{-5\text{ V}}{(8000\ \Omega)(0.00000003\text{ F})}$$
$$\approx 20{,}833\text{ V/s} = 20.833\text{ kV/s} \approx 21\text{ mV/μs}$$
$$\Delta V_{out} = (\Delta t)\,(21\text{ mV/μs})$$
$$= (500\text{ μs})\,(21\text{ mV/μs}) = 10{,}500\text{ mV} = \mathbf{10.5V}$$

The waveform:

- When $V_{in} = 0 \rightarrow V_{out} = 0$
- When $V_{in} = +5$ V $\rightarrow$ capacitor charging $\rightarrow V_{out} = \uparrow$

(-10.5 V $\rightarrow$ a negative going ramp)

– When $V_{in} = -5$ V → capacitor discharging → $V_{out} = \downarrow$
$$(+10.5 \text{ V} \rightarrow \text{a positive going ramp})$$

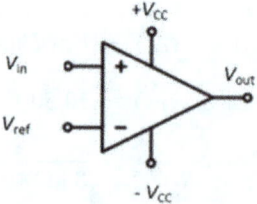

Figure 9.38   (a) Input wave / (b) Output wave

The output voltage of an integrator

$$V_{out} = \frac{-1}{RC} \int v_{in} dt$$

Derive: $\dfrac{\Delta V_{out}}{\Delta t} = - \dfrac{V_{in}}{R_{in} C}$

$\Delta V_{out} = - \dfrac{V_{in}}{R_{in} C} \Delta t$

or $d V_{out} = - \dfrac{V_{in}}{R_{in} C} dt,$ $V_{out} = \dfrac{-1}{R_{in} C} \int v_{in} dt$ Integrating both sides

9.3.5   Op-amp comparator

Introduction to comparator

- Op-amp comparator (voltage-level detector): an op-amp circuit configuration compares two input voltages (V_{in} and V_{ref}) and determines which is larger.
- Op-amp comparator circuit: Input → (+)

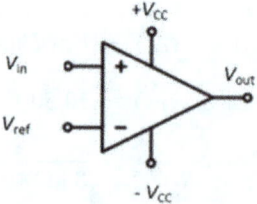

Figure 9.39   Comparator

- Compare voltages:
  - Any input voltage V_{in} above the reference voltage V_{ref} will produce a positive saturated output ($+V_{CC}$).
    If $V_{in} > V_{ref}$: $V_{out} = +V_{CC}$

- Any input voltage V_{in} below the reference voltage V_{ref} will produce a nega-tive saturated output $(-V_{CC})$.

  If $V_{in} < V_{ref}$: $V_{out} = -V_{CC}$
- A comparator used in an analog-to-digital conversion:

  A comparison of the two voltage levels V_{in} and V_{ref} can determine the digital output state, either a "1" or a "0."

  - From t_0 to t_1: $V_{in} < V_{ref} \rightarrow V_{out} = -V_{CC} \rightarrow$ "0"
  - From t_1 to t_2: $V_{in} > V_{ref} \rightarrow V_{out} = +V_{CC} \rightarrow$ "1"
  - From t_2 to t_3: $V_{in} < V_{ref} \rightarrow V_{out} = -V_{CC} \rightarrow$ "0"
  - ... ...

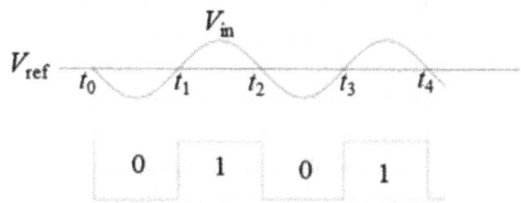

Figure 9.40   Analog to digital conversion

Summary

Operational amplifiers (op-amps)

- Operational amplifier (op-amp): an integrated circuit (IC) that operates as a voltage amplifier to amplify weak electric signals. It is a voltage-controlled amplifier with very high gain.
- Op-amp inputs:
  - Inverting input: the terminals labeled with the "+"
  - Noninverting input: the terminals labeled with the "–"

Table 9.4   Ideal op-amp vs. practical op-amp

Op Amp	A_v	Z_{in}	Z_{out}	BW	Circuit
Idea op-amp	$A_V = \infty$	$Z_{in} = \infty$	$Z_{out} = 0$	BW $= \infty$	
Practical op-amp	A_V is high	Z_{in} is high	Z_{out} is low	BW is limited	

The golden rules of op-amps

For an op-amp with external negative feedback:

- The current rule: no current flows into the two inputs of the op-amp.
$I_{in} = 0$
- The voltage rule: the potential difference between two inputs is zero in an op-amp circuit with negative feedback. $V_{in(-)} = V_{in(+)}$

Negative feedback

- Feedback: the process when all or a portion of the output signal is used as an input. It produces a path from the output back to the input.
- Positive feedback: when the occurrence of a reaction that enhances or amplifies the original action.
  - The output is connected to the noninverting input $V_{in(+)}$.
  - The fed-back signal is in phase with the input signal.
- Negative feedback: when the occurrence of a reaction reduces the original action and brings the system back to a stable state.
  - The output is connected to the inverting input $V_{in(-)}$.
  - The fed-back signal is out of phase with the input signal.

Negative feedback in op-amp

$$V_{out} \downarrow \;\rightarrow V_{in(-)} \downarrow \;\rightarrow (V_{in(+)} - V_{in(-)}) \uparrow$$
$$V_{out} \uparrow$$

Table 9.5   Input signal modes

Single-ended input	Double-ended input (differential input)	Common-mode input
- The input signal is applied to one of the inputs. - If the input is applied to the inverting input, the input and output signals are out of phase. - If input is applied to noninverting input, the input and output signals are in phase.	- Apply two opposite polarity signals to two inputs. - The output signal is a superimposing of two input signals.	- Two identical signals (noise) appear on both inputs. - In an ideal op-amp, the common-mode signal will be canceled out.

Table 9.6 Basic op-amp circuits

	Noninverting op-amp	Inverting op-amp	Voltage follower
Circuit	Input $\rightarrow$ (+)	Input $\rightarrow$ (−)	
Formulas	$A_{CL} = 1 + \dfrac{R_f}{R}$ $Z_i = Z_{in}\left(A_{OL}\dfrac{R_1}{R_1+R_f} + 1\right)$ $Z_o = \dfrac{Z_{out}}{1 + A_{OL}\dfrac{R_1}{R_1+R_f}}$	$A_{CL} = -\dfrac{R_f}{R_1}$ $Z_i = R_1$ $Z_o = \dfrac{Z_{out}}{1 + A_{OL}\dfrac{R_1}{R_1+R_f}}$	$A_{CL} = 1$ $Z_i = Z_{in}(A_{OL} + 1)$ $Z_o = \dfrac{Z_{out}}{1 + A_{OL}}$
Phase relationship	The input and output signals are in phase.	The input and output signals are out of phase.	The output "follows" the input signal.

Table 9.7 Glossary of op-amp key terms

Term	Definition
Differential input voltage (V_{diff})	The maximum voltage that can be applied between two input terminals of an op-amp without damaging it.
Open-loop voltage gain (A_{OL})	The voltage gain of the op-amp when no feedback is applied to the circuit.
Closed-loop voltage gain (A_{CL})	The voltage gain of the op-amp when negative feedback is applied to the circuit (closed-loop format).
Peak-to-peak output voltage swing	The maximum peak-to-peak output voltage that an op-amp physically provides at its output without waveform clipping.
Common-mode voltage range (CMVR) or input voltage range (IVR)	A range of input voltages that produces in the proper operation of the op-amp.
Common-mode gain (A_{CM})	The voltage gain for the common-mode voltage (noise) between the two inputs of the op-amp.
Common-mode signals (noise)	An identical signal that appears on both inputs of an op-amp.
Common-mode rejection	The ability of the op-amp to eliminate common-mode signals to both inputs from the output.
Common-mode rejection ratio (CMRR)	The measure of the ability of the op-amp to eliminate common-mode signals (noise). $- \quad \text{CMRR} = \dfrac{\text{open-loop gain}}{\text{common-mode gain}} = \dfrac{A_{OL}}{A_{CM}}$ $- \quad \text{CMRR} = 20 \log \left(\dfrac{A_{OL}}{A_{CM}}\right)$
Slew rate	The maximum rate of change in the output of an op-amp caused by a step change on the input (per unit of time).
Input offset voltage (V_{os})	The DC voltage applied to inputs of an op-amp that makes the output voltage exactly zero.
Input offset current (I_{os})	A difference in the input current that flows in or out of each of the input pins of an op-amp even if the output voltage is exactly zero.
Input bias current (I_{inB})	The DC current required by two input terminals of the op-amp with the output at a specified level (to properly operate the first stage). It is defined as the average of the two input currents. $I_{inB} = \dfrac{I_{in(+)} + I_{in(-)}}{2}$
Op-amp frequency response	The change in gain of an op-amp to respond to a change in the frequency of the input signal.

Table 9.8   Basic op-amp circuits

	Circuit	Output voltage
Summing amplifier		$$V_{out} = -\left(\frac{V_{in1}}{R_1} + \frac{V_{in2}}{R_2} + \frac{V_{in3}}{R_3} + \ldots \right) R_f$$
Difference amplifier		$$V_{out} = (V_{in2} - V_{in1}) \frac{R_2}{R_1}$$
Differentiator		$$V_{out} = -R_F C \frac{\Delta V_{in}}{\Delta t}$$ or $$V_{out} = -R_F C \frac{dV_{in}}{dt}$$
Integrator		— Rate of change of the output voltage: $$\frac{\Delta V_{out}}{\Delta t} = -\frac{V_{in}}{R_{in} C}$$ — Output voltage: $$V_{out} = \frac{-1}{RC} \int v_{in}\, dt$$
Comparator		If $V_{in} > V_{ref}$: $V_{out} = +V_{CC}$ If $V_{in} < V_{ref}$: $V_{out} = -V_{CC}$

Self-test

9.1

1. For an op-amp with external (   ) feedback, no current flows into the two inputs of the op-amp.
2. The potential difference between two inputs is (   ) in an op-amp circuit with negative feedback.
3. (   ) feedback can limit saturation and reduce distortion and noise.
4. The op-amp exhibits three modes of operation based on the kind of (   ).
5. If the input is applied to (   ) input, the input and output signals are out of phase.
6. Double-ended input amplifies the (   ) between both signals. The resultant output signal is a superimposing of two input signals.

9.2

7. Inverting op-amp is an op-amp circuit which inverting input (–) receives the input signal and feedback from the output of the op-amp. It has (   ) -phase output with respect to the input.
8. Determine the output voltage, input and output impedance for the circuit shown in Figure 9.41, if V_{in} is 40 mV, R_1 is 1 kΩ, R_f is 40 kΩ, Z_{out} is 60 Ω, and A_{OL} is 130,000.

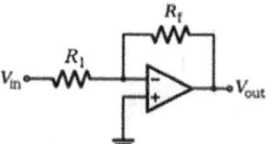

Figure 9.41   Ch 9: No. 8, self-test

9. Determine the output voltage, and input and output impedance for the circuit shown in Figure 9.42, if V_{in} is 40 mV, R_1 is 2 kΩ, R_f is 40 kΩ, A_{OL} is 130,000, Z_{in} is 1 MΩ and Z_{out} is 40 Ω.

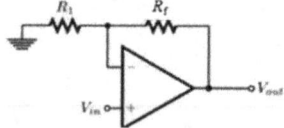

Figure 9.42   Ch 9: No. 9, self-test

10. The high input and low output impedances of the voltage follower make it a good "(   )" between two circuits.

11. Determine the output voltage, and input and output impedance for the circuit shown in Figure 9.43, if V_{in} is 30 mV, A_{OL} is 120,000, Z_{in} is 1.2 MΩ, and Z_{out} is 30 Ω.

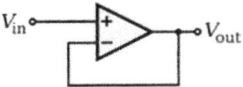

Figure 9.43   Ch 9: No. 11, self-test

12. The peak-to-peak output voltage swing is the maximum peak-to-peak output voltage that an op-amp physically provides at its output without waveform (   ).
13. The common-mode gain is the voltage gain for the common-mode voltage between the two (   ) of an op-amp.
14. The slew rate is the maximum rate of change in the output of an op-amp caused by a (   ) change on the input.
15. The input offset voltage is the DC voltage applied to inputs of an op-amp makes the output voltage exactly (   ).

9.3

16. The summing amplifier uses an (   ) amplifier configuration, and the noninverting input terminal is connected to ground.
17. If the resistors, R_1, R_2, R_3, R_f, etc. are all equal, a summing amplifier will have a (   ) gain.
18. Determine the output voltage for the circuit shown in Figure 9.44, if R_1 is 4 kΩ, R_2 is 2 kΩ, R_3 is 3 kΩ, R_f is 25 kΩ, V_{in1} is 4 mV, V_{in2} is 3 mV, and V_{in3} is 5 mV.

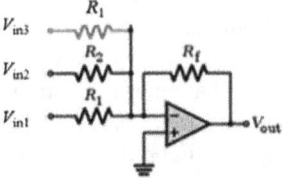

Figure 9.44   Ch 9: No. 18, self-test

19. Difference amplifiers are used mainly to reject (   ). Any signal common to both inputs will be automatically canceled out.
20. Difference amplifiers have high differential (   ) gain and very low common-mode gain.
21. Determine the output voltage for the difference amplifier circuit shown in Figure 9.45, if R_1 is 2 kΩ, R_2 is 3 kΩ, V_{in1} is 4mV, and V_{in2} is 5 mV.

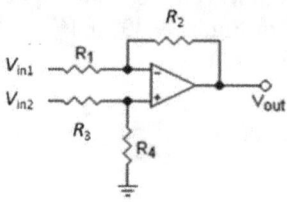

Figure 9.45   Ch 9: No. 21, self-test

22. An op-amp differentiator is basically an (   ) amplifier in which a resistor provides negative feedback and a capacitor at its input side.
23. If a triangular input signal is applied to a differentiator, the output will be changed to a (   ) wave.
24. Determine the output voltage for the differentiator shown in Figure 9.46, if R_f is 4 kΩ, C is 0.005 μF, and V_{in} is a triangular input signal rating from -10 V to $+10$ V at 15 μs.

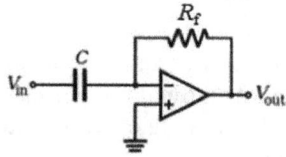

Figure 9.46   Ch 9: No. 24, self-test

25. An op-amp integrator produces an output that is proportional to the total (   ) included under the waveform.
26. If a square wave input signal is applied to an integrator, the output will be changed to a (   ) wave.
27. Determine the rate of change of the output voltage and the waveform for the integrator shown in Figure 9.47, if R_{in} is 10 kΩ, C is 0.04 μF, and V_{in} is a square wave input signal rating from -8V to $+8$V at 600 μs (the wave width). Assume that the voltage across the capacitor is initially zero.

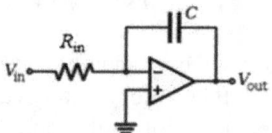

Figure 9.47   Ch 9: No. 27, self-test

Chapter 10

Oscillators

Chapter outline

10.1 Introduction to oscillators

10.1.1 Types of oscillators

Introduction to oscillators
- Oscillator: a device that produces a periodic sinusoidal signal or nonsinusoidal signal without an input AC signal (except a DC supply) as a source in signal generators.
- Oscillator vs. amplifier
  - Amplifier: an amplifier increases the strength of the input signal.
  - Oscillator: an oscillator generates a signal by itself without an input AC signal.

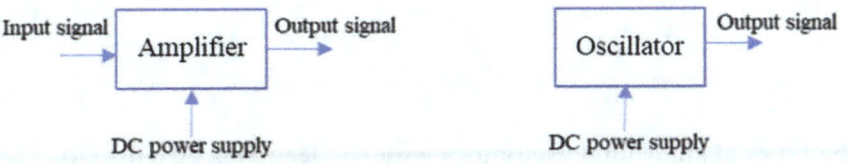

Figure 10.1   Oscillator vs. amplifier

- Output waveforms of oscillators: different types of oscillators can produce different types of output shapes, such as sine waves, square waves, triangular waves, sawtooth waves, etc.

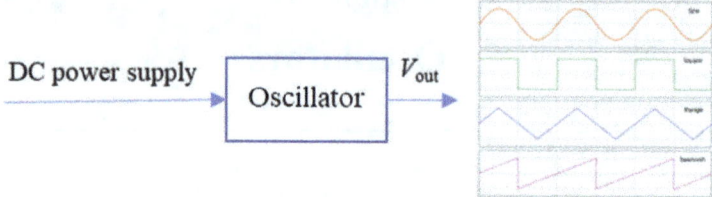

Figure 10.2   Oscillator waveforms

Feedback oscillators and relaxation oscillators

- There are two main types of oscillators: feedback oscillators and relaxation oscillators.
- Feedback oscillator (harmonic oscillators or linear oscillators): it is commonly used to generate sinusoidal waveforms.
- Relaxation oscillator: it is commonly used to generate nonsinusoidal waveforms including square waves, triangular waves, sawtooth waves, etc.

The types of nonsinusoidal oscillators (relaxation oscillators)

- Square wave oscillator
- Sawtooth wave oscillator (ramp oscillator)
- Triangular wave oscillator
- … …

  This book will only focus on sinusoidal oscillators.

The types of sinusoidal oscillators (feedback oscillators)

- There are two types of feedback oscillators: RC (resistor-capacitor) feedback oscillators and LC (inductor-capacitor) feedback oscillators.
- RC oscillator: a sinusoidal oscillator where an RC feedback circuit is used to give the required positive feedback and generate low-frequency signals.
- LC oscillator: a sinusoidal oscillator where an LC tuned circuit (or tank circuit) is used to give the required positive feedback and generate high-frequency signals.

  The LC oscillators are not practical to use at low frequencies (< 1MHz) because as the frequency becomes small, the value of the inductor (L) becomes large, the physical size of the inductor becomes large (large value inductors are bulky and costly). $f = \dfrac{1}{2\pi\sqrt{LC}}$

The types of sinusoidal oscillators with RC feedback circuits (RC oscillators)

- Wien-bridge oscillator (a popular type of audio frequency oscillator)
- Phase-shift oscillator
- Twin-T oscillator

The types of sinusoidal oscillators with LC feedback circuits (LC oscillators)

- Armstrong oscillator
- Colpitts oscillator
- Hartley oscillator
  They were named after their inventors.
- Cristal-controlled oscillator
- ...

Essential components in a sinusoidal oscillator:

- An amplifier: an amplifying device such as op-amp, bipolar junction transistor (BJT), or field-effect transistor (FET).
- A positive feedback circuit with a tuned or resonator circuit such as an LC circuit, or an RC circuit,etc.
  All oscillators use an amplifier with a positive feedback circuit and generate different operating frequencies.

10.1.2   Feedback and LC oscillators

Feedback

- There are two types of feedback: negative feedback and positive feedback
  - Negative feedback: a portion of the output is fed back to the input to reduce its output (the fed-back signal is out of phase with the input signal).

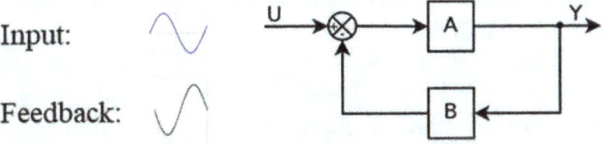

Figure 10.3   Negative feedback

  - Positive feedback: a portion of the output is fed back to the input to boost its output (the fed-back signal is in phase with the input signal).

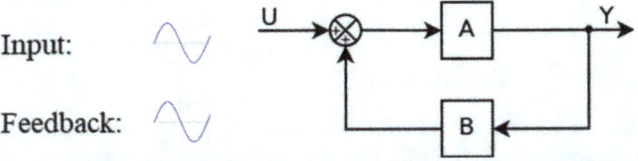

Figure 10.4   Positive feedback

The basic principle of oscillation

- Oscillators convert a DC source into an alternating output waveform at the desired frequency.
- The oscillation can be formed through the electric energy charging/discharging or storing/releasing among the capacitor and inductor.
- An LC oscillator uses an amplifier and positive feedback with a tuned or resonator circuit whose output is fed back to the input in phase. Thus, the signal sustains itself.

Parallel resonant circuit (tuned or tank circuit)

- Sinusoidal LC oscillators use a parallel LC resonant circuit to provide the oscillations.
- Resonance: the tendency to continue oscillating.
- Resonant circuit: the capacitor or inductor voltage/current in a resonant circuit could be much higher than the source voltage or current, a small input signal can produce a large output signal when resonance appears in a circuit.

Flywheel effect

- Flywheel effect in an oscillator: the ability of a tuned circuit to operate continuously after the control stimulus has been removed.
- The flywheel effect is produced by interacting capacitor and inductor in the tuned circuit of an oscillator.
  - When the power switch (BJT or FET) is on, the amplifier starts to conduct, the capacitor starts charging.
  - After the capacitor is fully charged, the power switch is off, the capacitor discharges through the inductor.
    When the capacitor gets discharged, the electric energy stored in the capacitor gets transferred in magnetic form to the inductor, and the inductor stores energy.
  - When the inductor starts to release energy, the capacitor gets charging again.
    When the switch is on: initial charging → C
    When the switch is off: C → discharges → L   L releases energy to C

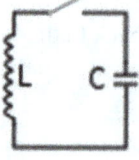

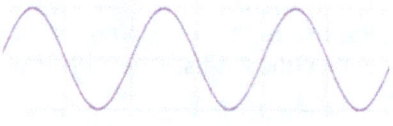

Figure 10.5b   Oscillation

Figure 10.5a   Tuned circuit

- This process of energy transfer between capacitor and inductor generates an oscillation in the resonant circuit (or tuned or tank circuit).
  This tank circuit can serve as an energy storage reservoir.

Damped oscillation

- Damped oscillation: an oscillation in which the amplitude of the oscillating quantity reduces over a period of time.
  The oscillator's amplitude continues to decrease with time.
- The cause of damped oscillations: there is small internal winding resistance in the inductor, or wires, etc. which dissipates energy over time.

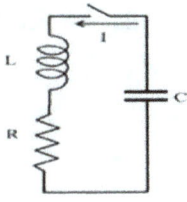

Figure 10.6(a)   R in the inductor   *Figure 10.6(b)   Damped oscillation*

- To replace the lost energy: turn on the power switch (amplifier) and the capacitor begins to charge again.

The conditions required for sustained oscillation

- Requirements for sinusoidal oscillation (Barkhausen criteria): an oscillator requires an amplifier and a positive feedback circuit with the following conditions:
  - the overall closed-loop phase shift equal to 0.
  - the overall closed-loop gain equal to 1 (unit gain, $A_v = 1$).
- If the gain is greater than 1, then eventually the output of the oscillator will become saturated.
  $A_v > 1 \rightarrow$ clipped
- If the gain is less than 1, the oscillations will eventually dampen out.
  $A_v < 1 \rightarrow$ die out

The initial condition for oscillation to start

- To start oscillation, the initial voltage gain of the amplifier must be greater than 1 ($A_v > 1$) so that a small signal is sufficient to trigger the amplifier, and the output can build up to the desired level.
- The gain must then reduce to 1 ($A_v = 1$) so that the output stays at the required level and oscillation is continued.

10.2   LC oscillators

10.2.1 Armstrong oscillator

Introductory to Armstrong oscillator

- Armstrong oscillator (or Meissner oscillator): the earliest oscillator that uses a transformer and capacitor in its feedback path to produce a sinusoidal output waveform at the desired frequency.
- The Armstrong oscillator is generally used as an oscillator in receivers.

A basic Armstrong oscillator circuit

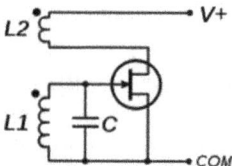

Figure 10.7   Armstrong oscillator

- The inductance of the transformer winding (inductor coil), along with a capacitance, forms a resonant circuit of the Armstrong oscillator.
- Positive feedback is accomplished by the tickler coil L_2 and the L_1C circuit (mutual inductance in transformer).
- Tickler coil: a small coil in series with the plate circuit and coupled to the grid to provide feedback.

Oscillation frequency

- The frequency of the Armstrong oscillator can be controlled by the tuned LC circuit (L_1 and C).
- Oscillation frequency:

$$f \approx \frac{1}{2\pi \sqrt{L_1 C}}$$

  - L_1: the primary inductance of the transformer.
  - C: the capacitance of the tank circuit.

10.2.2 Colpitts oscillator

Introduction to Colpitts oscillator

- Colpitts oscillator: an LC oscillator that uses an inductor in parallel with two capacitors in series (or a single center-tapped capacitor) in the feedback path to produce a sinusoidal output waveform at the desired frequency.

- The positive feedback is taken from the center connection of the two capacitors (or the center tap of the capacitor) of the resonant or tank circuit.

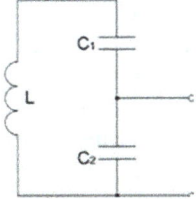

Figure 10.8   LC circuit

A basic Colpitts oscillator circuit

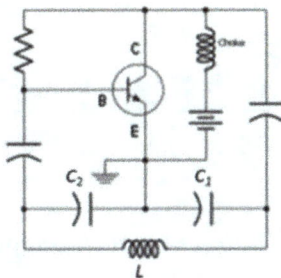

Figure 10.9   A basic Colpitts oscillator

Oscillation frequency of Colpitts oscillator
- The frequency of oscillation can be determined by using the values of the capacitor and inductor of the tank circuit.
- Oscillation frequency: $f = \dfrac{1}{2\pi \sqrt{LC_{eq}}}$

$$C_{eq} = \frac{C_1 C_2}{C_1 + C_2} \qquad \qquad C_1 \text{ and } C_2 \text{ are in series}$$

$$\text{If } C_1 = C_2 = C, \quad C_{eq} = \frac{C}{2}$$

- L: the inductance of the tank circuit.
- C_{eq}: the equivalent capacitance of two series-connected capacitors of the tank circuit.

    Recall: - two capacitors in series: $C_{eq} = \dfrac{C_1 C_2}{C_1 + C_2}$

    - two capacitors in parallel: $C_{eq} = C_1 + C_2$

- The frequency of oscillation can be adjusted (tuned) by using a variable capacitor (varying the C) or by varying the position of the core inside the coil (varying the L).

10.2.3   Hartley oscillator

Introduction to Hartley oscillator

*   Hartley oscillator: an LC oscillator that uses a capacitor in parallel with two inductors in series (or a single center-tapped inductor) in the feedback path to produce a sinusoidal output waveform at the desired frequency.
*   The positive feedback is taken from the center connection of the two inductors (or the center tap of the inductor) of the resonant or tank circuit.

A basic Hartley oscillator circuit

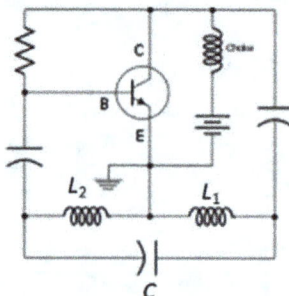

Figure 10.10   A basic Hartley oscillator

The main difference between a Colpitts oscillator and a Hartley oscillator

*   Colpitts
    *   The resonant circuit uses an inductor and two capacitors (or a center-tapped capacitor).
    *   The feedback signal is taken from the center tap of the two capacitors.
*   Hartley
    *   The resonant circuit uses one capacitor and two inductors (or a center-tapped inductor).
    *   The feedback signal is taken from the center tap of the two inductors.

Oscillation frequency of Hartley oscillator

*   The frequency of oscillation can be determined by using the values of the capacitor and inductor of the tank circuit.
*   Oscillation frequency: $f = \dfrac{1}{2\pi\sqrt{L_T C}}$

* $L_T = L_1 + L_2$
    * C: the capacitance of the tank circuit.
    * L_T: the total inductance of two series-connected inductors of the tank circuit.
    Recall: two inductors in series: $L_T = L_1 + L_2$

10.2.4 Cristal-controlled oscillator

Introduction to the crystal-controlled oscillator

- Crystal-controlled oscillator: an LC oscillator that uses the piezoelectric crystal to obtain a piezoelectric effect in the feedback path to produce a sinusoidal output waveform at the desired frequency.
- The crystal-controlled oscillator is the most stable and accurate type of feedback oscillator.

Piezoelectric effect

- Piezoelectric effect: the ability of certain crystals (quartz, Rochelle salt, etc.) to accumulate electric charges and generate a voltage at the frequency of vibration in the material when pressure is applied to it.
  The term "piezo" is derived from the Greek, which means to press, push, squeeze, etc.
- Conversely, when an AC voltage is applied across the crystal, it vibrates at the frequency of the applied voltage.
- Symbol of piezoelectric crystal:

Figure 10.11   Symbol of piezoelectric crystal

A piezoelectric sandwich: a nonconductive piezoelectric crystal is placed between the two metal plates.

Equivalent circuit for a piezoelectric crystal

- The equivalent circuit of a piezoelectric crystal is represented by a series RLC circuit in parallel with a capacitor C_p.
- Equivalent model:

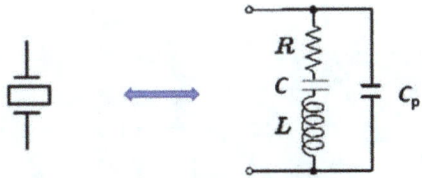

Figure 10.12   Equivalent circuit of a piezoelectric crystal

Oscillation frequency

- The greatest vibration occurs at the natural resonant frequency of the crystal.
- The frequency of oscillation can be determined by the way the crystal is cut or the physical size of the crystal.

Crystal-controlled oscillator circuits

- The crystal-controlled oscillator can operate in either series resonance or parallel resonance.
- Parallel resonant mode: the impedance of the crystal is maximum, and the voltage is maximum at the parallel resonant frequency to provide the largest positive feedback.

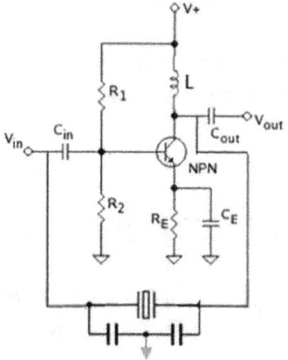

Figure 10.13   Parallel resonant mode

- Series resonant mode: the impedance of the crystal is minimum, and the current is maximum at the series resonant frequency to provide the largest positive feedback.

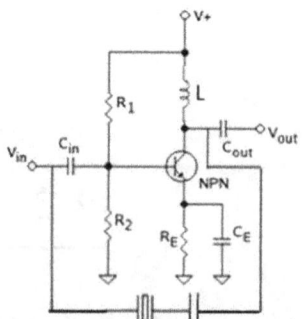

Figure 10.14   Series resonant mode

10.3 RC oscillators

RC oscillators (sinusoidal oscillators with RC feedback circuits)
- An RC oscillator uses a positive-feedback loop that consists of an amplifier and an RC frequency selecting circuit to generate a low-frequency signal.
- There are three types of RC oscillators: Wien bridge, Twin-T, and phase-shift.
- Wien bridge oscillator is a popular low-frequency RC oscillator. It has good stability, constant output, and ease of operation (ease of tuning).

Op-amp Wien bridge oscillator
- Wien bridge oscillator: an RC oscillator that uses a Wien bridge to provide feedback with no phase shift and generate sinusoidal output waveforms at the desired frequency.
- The Wien bridge oscillator is also called a Wheatstone bridge circuit because it is based on the Wheatstone bridge theory. The bridge consists of four resistors and two capacitors.

Wien bridge oscillator circuit

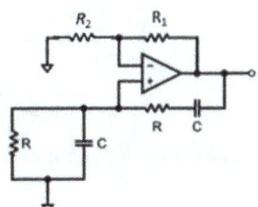

Figure 10.15   Wien bridge oscillator

- Four arms in the RC bridge circuit:
  - R_1 and R_2 (purely resistive) form the two arms of the bridge circuit.
  - One R and C are connected in parallel to form one arm of the bridge circuit.
  - One R and C are connected in series to form one arm of the bridge circuit.
- Frequency adjustment parts: the branches R and C are the frequency adjustment parts.

The oscillation frequency of the Wien bridge oscillator
- The frequency for a Wein bridge oscillator is calculated using the resonant frequency.
- Resonant frequency: the natural frequency of a system that tends to reach the maximum value (the phase shift is 0°).
- Oscillation frequency: $$f = \frac{1}{2\pi RC}$$

The resonant frequency in an RC circuit: $f = \frac{1}{2\pi RC}$

Gain of the Wien bridge oscillator

- Recall – requirements for oscillation: a feedback oscillator requires a positive feedback circuit and an op-amp with the overall gain equal to 1 (unit gain, $A_v = 1$).
- Positive-feedback path: the branches R_1 and R_2 in the Wien bridge oscillator are part of the positive-feedback path.
- Since there is a loss of about one-third ($\frac{1}{3}$) of the gain in the feedback path, the gain of the amplifying branch (op-amp) should be 3 so that the overall loop gain condition ($A_v = 1$) is satisfied.
- To start oscillation, the initial closed-loop gain of the op-amp must be greater than 3 ($A_{CL} > 3$) so that any small signal is sufficient to trigger the amplifier, and the output can build up to the desired level.
- The gain must then reduce to 3 ($A_{cl} = 3$) so that the overall gain around the loop is 1 ($A_V = 1$), and the output stays at the required level and oscillation is continued.

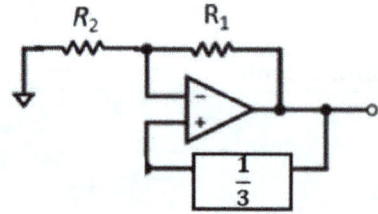

Figure 10.16   Gain of the Wien bridge oscillator

- The ratio of resistors R_1 and R_2 in the positive feedback path determines the gain of the Wien bridge oscillator.
- To achieve the required level gain condition: resistance R_1 should be twice the value of R_2.
- $R_1 = 2R_2$

Example: For the Wien bridge oscillator of Figure 10.15, determine the value of capacitance C and resistance R_2, if resistance R is 50 kΩ, R_1 is 200 kΩ, and the oscillation frequency is 5,000 Hz.

- Given: $R = 50\ k\Omega$, $R_1 = 200\ k\Omega$, and $f = 5,000$ Hz.
- Find: C and R_2.

- Solution: $C = \dfrac{1}{2\pi R f} = \dfrac{1}{2\pi\ (50\ k\Omega)(5,000\ \text{Hz})}$

    $= 6.36 \times 10^{-10}$ F = **636 pF**   Kilo: 10^3; Pico: 10^{-12}

    $R_2 = \dfrac{R_1}{2}\ \dfrac{200\ k\Omega}{2} = $ **100 kΩ** $R_1 = 2R_2$

The phase shift of the Wien bridge oscillator

*   Recall – requirements for oscillation: a feedback oscillator requires a positive feedback circuit with zero-phase shift.
*   In a general oscillator, an amplifier produces a180° phase shift, and a positive feedback path produces an additional 180° phase shift to obtain the total zero-phase shift to satisfy the conditions for oscillation (Barkhausen criteria).
*   The Wien bridge oscillator provides a phase shift of 0° in the feedback path because it uses a non inverting amplifier (it does not produce phase shift).

Summary

Oscillators

*   Oscillator: a device that produces a periodic sinusoidal signal or nonsinusoidal signal without an input AC signal (except a DC supply) as a source in signal generators.
*   Oscillator vs. amplifier
    *   Amplifier: an amplifier increases the strength of the input signal.
    *   Oscillator: an oscillator generates a signal by itself without an input AC signal.

Types of oscillators

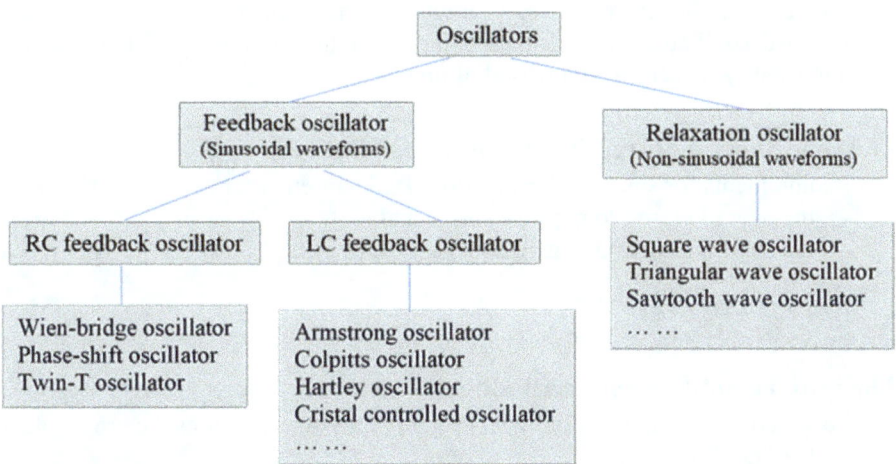

Essential components in a sinusoidal oscillator:

*   An amplifier: an amplifying device such as op-amp, bipolar junction transistor (BJT), or field-effect transistor (FET).
*   A positive-feedback circuit with a tuned or resonator circuit such as an inductor-capacitor (LC) circuit, or a resistor-capacitor (RC) circuit, etc.

The basic principle of oscillation

- Oscillators convert a DC source into an alternating output waveform at the desired frequency.
- The oscillation can be formed through the electric energy charging/discharging or storing/releasing among the capacitor and inductor.
- An LC oscillator uses an amplifier and positive feedback with a tuned or resonator circuit whose output is fed back to the input in phase. Thus, the signal sustains itself.

Parallel resonant circuit (tuned or tank circuit)

- Sinusoidal LC oscillators use a parallel LC resonant circuit to provide the oscillations.
- Resonant circuit: the capacitor or inductor voltage/current in a resonant circuit could be much higher than the source voltage or current, a small input signal can produce a large output signal when resonance appears in a circuit.

Flywheel effect

- Flywheel effect in an oscillator: the ability of a tuned circuit to operate continuously after the control stimulus has been removed.
- The flywheel effect is produced by interacting capacitor and inductor in the tuned circuit of an oscillator.
- This process of energy transfer between capacitor and inductor generates an oscillation in the resonant circuit (or tuned or tank circuit).
- Damped oscillation: an oscillation in which the amplitude of the oscillating quantity reduces over a period of time.

The conditions required for sustained oscillation

- Requirements for sinusoidal oscillation (Barkhausen criteria):
  - the overall closed-loop phase shift equal to 0.
  - the overall closed-loop gain equal to 1 (unit gain, $A_v = 1$).
- If $A_v > 1$ → clipped
- If $A_v < 1$ → die out

The initial condition for oscillation to start

- To start oscillation, the initial voltage gain of the amplifier must be greater than 1 ($A_v > 1$).
- The gain must then reduce to 1 ($A_v = 1$) so that the output stays at the required level, and oscillation is continued.

Table 10.1 Basic oscillators

Oscillator	Basic circuit	Oscillation frequency	Information
Armstrong oscillator	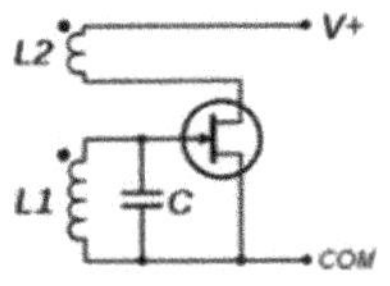	$f \approx \dfrac{1}{2\pi\sqrt{L_1 C}}$	- L_1: the primary inductance of the transformer. - C: the capacitance of the tank circuit.
Colpitts oscillator	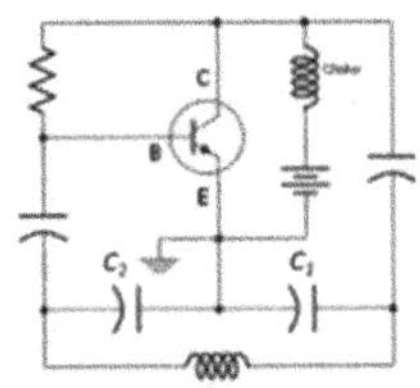	$f \approx \dfrac{1}{2\pi\sqrt{LC_{eq}}}$ $C_{eq} = \dfrac{c_1 c_2}{c_1 + c_2}$ If $C_1 = C_2 = C$ $C_{eq} = \dfrac{c}{2}$	- L: the inductance of the tank circuit. - C_{eq}: the equivalent capacitance of two series-connected capacitors of the tank circuit.
Hartley oscillator	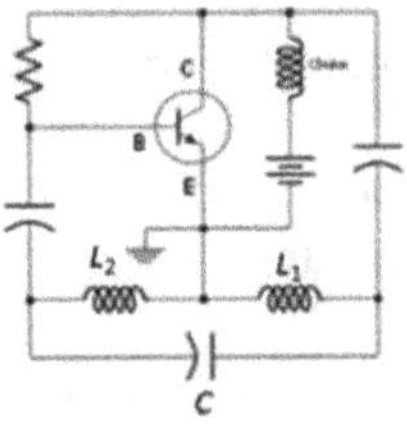	$f \approx \dfrac{1}{2\pi\sqrt{L_T c}}$ $L_T = L_1 + L_2$	- C: the capacitance of the tank circuit. - L_T: the total inductance of two series-connected inductors of the tank circuit.

Crystal-controlled oscillator

Parallel resonant mode	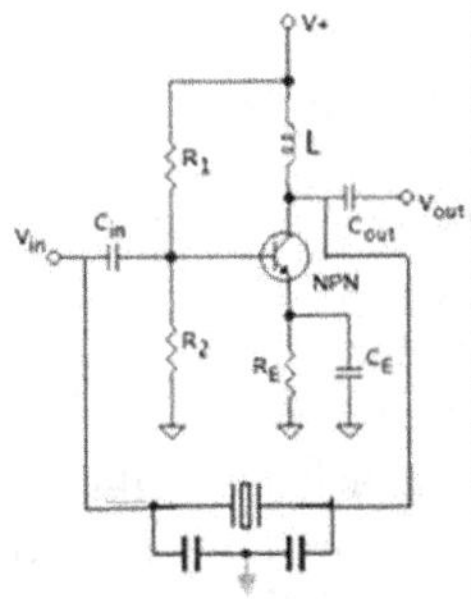	- The frequency of oscillation can be determined by the way the crystal is cut or the physical size of the crystal. - The greatest vibration occurs at the natural resonant frequency of the crystal.	Equivalent circuit for a piezoelectric crystal: 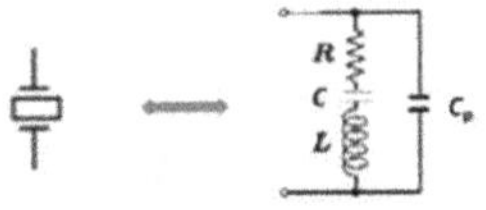
Series resonant mode	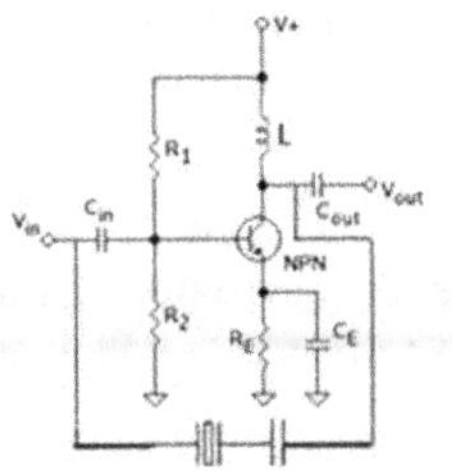		

(Continued)

Table 10.1   Basic oscillators (Continued)

Oscillator	Basic circuit	Oscillation frequency	Information
Wien bridge oscillator		$f \approx \dfrac{1}{2\pi RC}$	To achieve the required level gain condition: $R_2 = 2R_1$

Self-test

10.1

1. An oscillator generates a signal by itself without an input (   ) signal.
2. The (   ) oscillator is commonly used to generate sinusoidal waveforms.
3. The (   ) oscillator is commonly used to generate nonsinusoidal waveforms including square waves, triangular waves, sawtooth waves, etc.
4. The RC oscillator is a sinusoidal oscillator where an RC feedback circuit is used to give the required positive feedback and generate (   ) -frequency signals.
5. The LC oscillator: a sinusoidal oscillator where an LC (   ) or tank circuit is used to give the required positive feedback and generate high-frequency signals.
6. A (   ) feedback circuit with a tuned or resonant circuit such as an LC circuit, or an RC circuit, etc.
7. An LC oscillator uses an amplifier and (   ) feedback with a tuned or resonant circuit whose output is fed back to the input in phase.
8. The (   ) effect in an oscillator is the ability of a tuned circuit to operate continuously after the control stimulus has been removed.
9. The (   ) oscillation is an oscillation in which the amplitude of the oscillation quantity reduces over a period of time.
10. An oscillator requires an amplifier and a positive feedback circuit with the overall closed-loop gain equal to (   ).

10.2

11. An Armstrong oscillator is the earliest oscillator that uses a (   ) and capacitor in its feedback path to produce a sinusoidal output waveform at the desired frequency.
12. The (   ) coil is a small coil in series with the plate circuit and coupled to the grid to provide feedback.

13. The frequency of the Armstrong oscillator can be controlled by the tuned (   ) circuit.

14. A Colpitts oscillator is an LC oscillator that uses an (   ) in parallel with two capacitors in series (or a single center-tapped capacitor) in the feedback path to produce a sinusoidal output waveform at the desired frequency.

15. The frequency of oscillation can be determined by using the values of the capacitor and inductor of the (   ) circuit.

16. The Hartley oscillator is an LC oscillator that uses a (   ) in parallel with two inductors in series (or a single center-tapped inductor) in the feedback path to produce a sinusoidal output waveform at the desired frequency.

17. A crystal-controlled oscillator is an LC oscillator that uses the piezoelectric crystal to obtain a (   ) effect in the feedback path to produce a sinusoidal output waveform at the desired frequency.

18. The (   ) effect is the ability of certain crystals to accumulate electric charges and generate a voltage at the frequency of vibration in the material when pressure is applied to it.

19. The equivalent circuit of a piezoelectric crystal is represented by a series RLC circuit in parallel with a (   ).

10.3

20. The Wien bridge oscillator is an RC oscillator that uses a Wien bridge to provide (   ) with no phase shift and generate sinusoidal output waveforms at the desired frequency.

21. The frequency for a Wein bridge oscillator is calculated using the (   ) frequency.

22. The ratio of resistors R_1 and R_2 in the positive feedback path determines the (   ) of the Wien bridge oscillator.

23. For the Wien bridge oscillator of Figure 10.17, determine the value of capacitance C and resistance R_2, if resistance R is 40 kΩ, R_1 is 100 kΩ, and the oscillation frequency is 4,000 Hz.

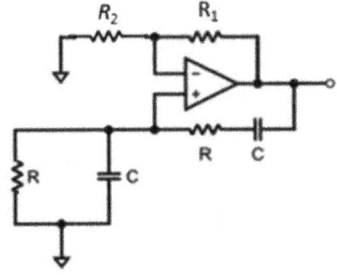

Figure 10.17   Ch 10: No. 23, self-test

Chapter 11

Voltage regulators

Chapter outline

11.1   Introduction to op-amp voltage regulators

11.1.1   Voltage regulators- concepts review

Power supply

- Voltage regulator: a device that is designed to automatically maintain a stable output voltage despite variations in input signal or load.
- DC voltage source: a battery is the most common DC voltage source, and a power supply is another type of the DC voltage source.
- DC power supply: a device that provides DC power to an electrical/electronic equipment. It can convert the AC power from an electrical wall outlet to a steady DC output.
- One method of obtaining DC power supply is to transform, rectify, filter, and regulate an AC voltage (each of that performs a specific function).
- The stages of a power supply:
  - Transformer: a device that transfers AC power from input to output, either increasing (stepping up) or reducing (stepping down) the input AC voltage according to the requirement.

- Rectifier: a device that converts two-directional AC voltage (sine waves) into single-directional DC (pulsating DC).
- Filter: a device that smooths the voltage waveform at the output of the rectifier from varying greatly to a small ripple.
- Voltage regulator: a device that eliminates ripple at the output of the filter by setting output to a constant DC voltage.

Figure 11.1   DC power supply

- Types of voltage regulators: there are two types of voltage regulators– linear voltage regulator and switching voltage regulator.

Line regulation and load regulation

- Line regulation: the ability of a voltage regulator to maintain the output voltage V_{out} level with a varying input voltage V_{in}.

$$V_{in} \rightarrow V_{out}$$

- Load regulation: the ability of a voltage regulator to maintain the output voltage V_{out} level with a varying load.

$$R_{L} \rightarrow V_{out}$$

Percent regulation

- Percent regulation can be used to specify the performance of a voltage regulator. It can be in terms of line regulation or load regulation.

- Percent line regulation is expressed as the percent of a small change in the output voltage over a small change in the input voltage.

Percent line regulation: $\left(\dfrac{\Delta V_{output}}{\Delta V_{input}} \right) 100\%$ Δ means "a change in"

- Percent load regulation is not a fixed number but rather presented as a percentage in response to changes at the output. It means over the permissible load range (from minimum or no load to full load) the regulation can change.

Percent load regulation: $\left(\dfrac{V_{No\ load} - V_{Full\ load}}{V_{Full\ load}} \right) 100\%$

The smaller % regulation, the better.

- $V_{No\ load}$: percent voltage with no load (or minimum load).
- $V_{Full\ load}$: percent voltage with a full load.

11.1.2 Classification of op-amp voltage regulators

Classification of voltage regulators

- Types of voltage regulators: voltage regulators can be generally classified into two types:
  - Linear voltage regulators
  - Switching voltage regulators
- Linear voltage regulator: a voltage regulator that a linear component (such as a transistor stays in the active/linear region of its operation) during voltage regulation is used to regulate the output voltage.

- Switching voltage regulator: a voltage regulator that a switching component (such as a transistor operates in cut-off or saturation state) during voltage regulation is used to regulate the output voltage.

- Efficiency: switching regulators operate with higher power-handling capacity and efficiency than linear regulators because the transistor is not always conducting.

Classification of linear voltage regulators

- Types of linear voltage regulators: linear voltage regulators can be classified into two types:
  - Series voltage regulators
  - Shunt voltage regulators
- Series voltage regulator: a voltage regulator that is connected in series with the load.

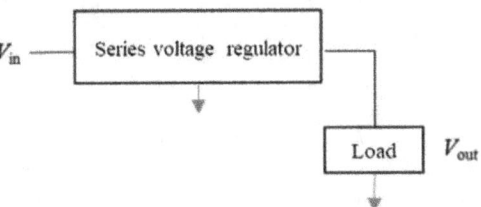

Figure 11.2   Series voltage regulator mode

- Shunt voltage regulator: a voltage regulator that is connected in parallel with the load.

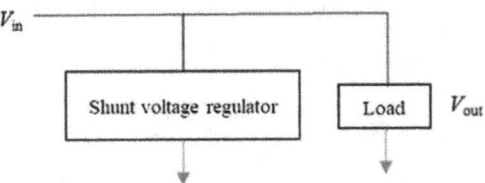

Figure 11.3   Shunt voltage regulator mode

Classification of switching voltage regulators

- Types of switching voltage regulators: switching voltage regulators can be classified into three types:
  - Step-down (buck) voltage regulators
  - Step-up (boost) voltage regulators
  - Inverting (flyback) voltage regulators
- Step-down voltage regulator: a switch-mode voltage regulator that can convert an input voltage to a lower output voltage ($V_{out} < V_{in}$).

  The power switch is in series with the input and the inductor.

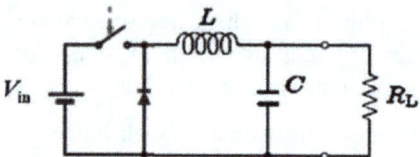

Figure 11.4   Step-down voltage regulator

- Boost (step-up) voltage regulator: a switch-mode voltage regulator that can convert an input voltage to a higher output voltage ($V_{out} > V_{in}$).

  The power switch is after the inductor and in parallel with the input and the inductor branch.

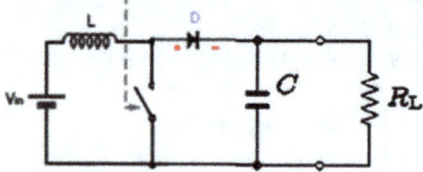

Figure 11.5   Step-up voltage regulator

- Inverting voltage regulator: a switch-mode voltage regulator that output voltage is opposite polarity of the input signal (V_{out} and V_{in} out of phase).

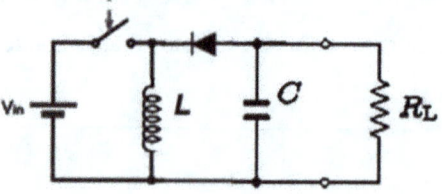

Figure 11.6   Inverting voltage regulator

The power switch is in series with the input and in parallel with the inductor.

Basic components in a voltage regulator

- Reference element (Zener diode): sets a reference voltage V_{ref} (it is the voltage across the Zener diode V_z).
- Control element (BJT or MOSFET): delivers the required current (compensates) to keep a constant output voltage V_{out}.
- Resistive voltage divider (R_1 and R_2): provides a sample of output voltage for feedback.
- Error detector (op-amp comparator): compares the feedback output voltage V_{out} with a reference voltage V_{ref}.

11.2 Linear op-amp voltage regulators

11.2.1 Series voltage regulators

A basic series voltage regulator circuit

- Series voltage regulator (or series-pass voltage regulator): a voltage regulator that uses a control element (series-pass transistor) connected in series with the load.
- The circuit of a basic series voltage regulator:

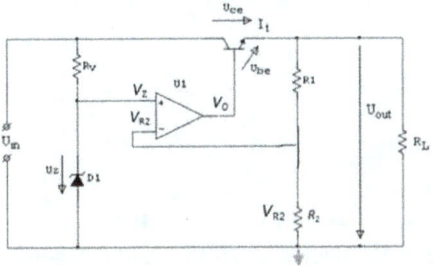

Figure 11.7   Series voltage regulator

Series voltage regulator basics

- The resistors R_1 and R_2 (voltage divider) help in sensing the variations in the output and provide a feedback signal to op-amp (inverting input).
- The op-amp compares the V_{ref} (or V_z) and V_{out} (or V_{R2}) and adjusts the base current of the transistor to deliver the required load current while maintaining a constant output voltage.
- Transistor: depending on the resulting difference voltage (feedback and the reference), a control signal is then produced to drive the transistor to compensate for the variations.
  A transistor can add current drive capacities.
- Constant output voltage: as a result, the voltage across the transistor is varied, and the output voltage will be retained at a steady value.

The output voltage of a series voltage regulator

- Output voltage: $V_{out} \approx V_Z \left(1 + \dfrac{R_1}{R_2}\right)$

Derive: $V_{R2} = V_{out} \dfrac{R_2}{R_1 + R_2}$ Voltage divider rule

$V_{R2} \approx V_Z$ Golden rules of op-amps (voltage rule)

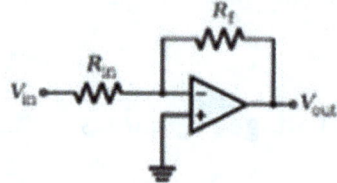

Figure 11.8   The voltage rule

$V_Z = V_{out} \dfrac{R_2}{R_1 + R_2}$

$V_{out} = V_Z \dfrac{R_1 + R_2}{R_2}$

$\therefore V_{out} = V_Z \left(\dfrac{R_1}{R_2} + 1\right)$

Negative feedback in series voltage regulator

- If the output voltage V_{out} decreases (if the load resistance R_{Load} or input voltage V_{in} changed), then the voltage across R_2 (V_{R2}) will fall by the same amount so the difference in the inputs ($V_Z - V_{R2}$) will increase.

$$V_{out}\!\downarrow = V_{R2}\!\downarrow \quad \rightarrow \quad (V_Z - V_{R2})\!\uparrow$$

- This causes the transistor conduct more current and output voltage V_{out} to rise:

$$A_V(V_Z - V_{R2})\!\uparrow \rightarrow V_O\!\uparrow \quad \rightarrow \quad I_B\!\uparrow \quad \rightarrow \quad I_C\!\uparrow \quad \rightarrow \quad V_{out}\!\uparrow$$

  − I_B and I_C: the base and collect currents of the BJT.
  − V_O: op-amp output voltage, $A_V = \dfrac{V_{out}}{V_{in}}$, $V_O = A_V(V_Z - V_{R2})$

$$V_O\!\uparrow \quad \rightarrow \quad V_{out}\!\uparrow$$

- Negative feedback in the series op-amp voltage regulator:

$$V_{out}\!\downarrow \rightarrow V_{R2}\!\downarrow \rightarrow (V_Z - V_{R2})\!\uparrow \quad \rightarrow \quad V_O\!\uparrow \quad \rightarrow \quad I_B\!\uparrow \quad \rightarrow \quad I_C\!\uparrow \qquad I_C = I_E$$
$$V_{out}\!\uparrow \longleftarrow$$

The op-amp will adjust its output to keep the voltage difference between two inputs zero when feedback is being used.

Series voltage regulator – working principle

- The op-amp comparator circuit tries to make V_Z approximately equal to V_{R2} all the time.

$$V_{in} \text{ or } R_{Load} \rightarrow V_{out} \rightarrow V_{R2} \rightarrow V_{R2} = (V_Z)$$

$$V_{out} \longleftarrow$$

The golden rule of op-amps (voltage rule)

- The output voltage V_{out} is independent of the load resistor and input voltage V_{in}, so a relatively constant output voltage is maintained.

$$V_{out} = V_Z \left(1 + \frac{R_1}{R_2} \right)$$

11.2.2   Shunt voltage regulators

A basic shunt voltage regulator circuit

- A basic shunt voltage regulator: a voltage regulator that uses a control element (transistor) connected in parallel with the load.
- The circuit of a basic shunt voltage regulator:

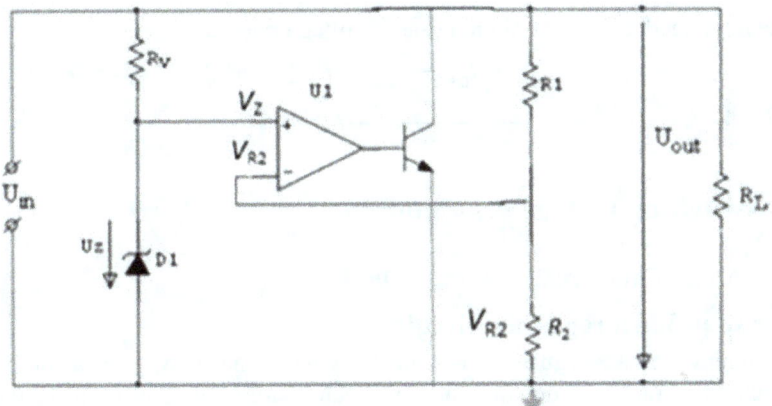

Figure 11.9   Shunt voltage regulator

- The shunt voltage regulator regulates output voltage by shunting current away (to ground) from the load to regulate the output voltage.
- The function of the shunt voltage regulator is similar to that of the series voltage regulator except that regulation is attained by controlling the current through the parallel transistor.

Shunt voltage regulator basics

- The resistors R_1 and R_2 (voltage divider) help in sensing the variations in the output and provide a feedback signal to the op-amp.
- The op-amp is used as a comparator that compares the feedback voltage with a reference voltage (the voltage across the Zener diode V_z).
- Transistor: some of the current is shunted away by the control element transistor.
- Constant output voltage: the resulting difference voltage provides a control signal to change the amount of the current shunted away from the load, and the output voltage will be retained at a steady value.

Negative feedback in shunt voltage regulator

- If output voltage V_{out} decreases (when the load resistance R_{Load} or input voltage V_{in} changed), then the voltage across R_2 (V_{R2}) will fall by the same amount.
- This causes the transistor to conduct less base current (I_B) and increases the voltage at the collector and emit (V_{CE}).

$$V_{out}\downarrow \;\; \rightarrow V_{R2}\downarrow \; \rightarrow \; I_B\downarrow \rightarrow I_C\downarrow \; \rightarrow \; V_{CE}\uparrow \qquad\qquad I_B = \beta I_C, V_{CE} = V_{CC} - I_C R_C$$

- The transistor produces more voltage, and the output voltage V_{out} also raises so that a constant output voltage is maintained.

$$V_{CE}\uparrow \rightarrow V_{out}\uparrow \qquad\qquad V_{CE} = V_{out}$$

- Negative feedback in the shunt op-amp voltage regulator:

$$V_{out}\downarrow \rightarrow V_{R2}\downarrow \rightarrow V_O\downarrow \rightarrow I_B\downarrow \rightarrow I_C\downarrow \rightarrow V_{CE}\uparrow \qquad V_O\text{: op-amp output voltage}$$

$$V_{out}\uparrow \; \longleftarrow \underline{\hspace{6cm}}$$

11.3 Switching voltage regulators

11.3.1 Step-down voltage regulator

A basic step-down regulator circuit

- Switching voltage regulator: a voltage regulator that a switching component (such as a transistor operates in cut-off and saturation state) during voltage regulation is used to regulate the output voltage.
- Step-down voltage regulator: a switch-mode voltage regulator that can convert an input voltage to a lower output voltage ($V_{out} < V_{in}$).
- Simplified step-down regulator diagram: the transistor (switching component) is represented by a switch driven by an error amp.

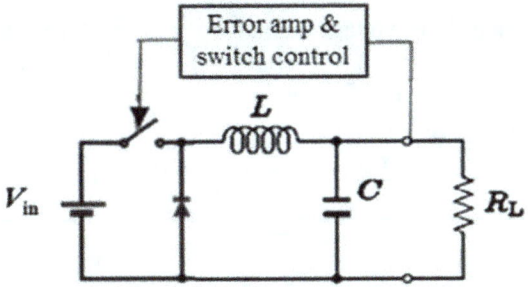

Figure 11.10 Step-down regulator

Step-down voltage regulator – working principle
- When the power switch (transistor) is on:
  - The diode is reverse biased (RB) (diode is off).
  - The capacitor charges (the longer the switch is on, the greater the V_{out}).
- When the power switch (transistor) is off:
  - The diode is forward biased (FB) (diode is on).
  - The capacitor discharges to the load.

Waveforms of a step-down voltage regulator:

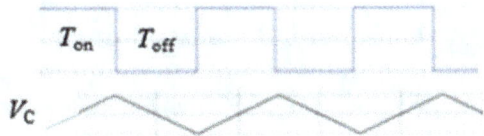

Figure 11.11 Waveforms of a step-down voltage regulator

- On and off time:
  - On time ton: the time that transistor devotes in saturation (switch is on, capacitor charges).
  - Off time toff: the time that transistor devotes in cutoff (switch is off, capacitor discharges).
- Output voltage and period:
  - Output voltage:
  $$V_{out} = V_{in} \left(\frac{t_{on}}{t_{on} + t_{off}} \right)$$

  - Period of the on-and-off cycle: $\quad T = t_{on} + t_{off}$
- Repeating waveform: the waveform repeats itself with the repetition on and off action of the switch.
- Negative feedback in the step-down voltage regulator:

$$V_{out}\downarrow \quad \longrightarrow \quad t_{on}\uparrow$$
$$V_{out}\uparrow \quad \longleftarrow \quad \rfloor$$

$$V_{out} = V_{in} \left(\frac{t_{on}}{t_{on} + t_{off}} \right)$$

Example: For the step-down voltage regulator of Figure 11.10, determine the output voltage, if the input voltage is 12 V, on time is 2 μs, and off time is 4 μs.

- Given: $V_{in} = 12$ V, $t_{on} = 2$μs, and $t_{off} = 4$ μs.
- Find: V_{out}

- Solution: $V_{out} = V_{in} \left(\dfrac{t_{on}}{t_{on} + t_{off}} \right) = 12$ V $\left(\dfrac{2 \ \mu s}{2 \ \mu s + 4 \ \mu s} \right) = \textbf{4V}$ Step-down:$12V \rightarrow 4$ V

11.3.2 Step-up voltage regulator

A basic step-up regulator circuit

- Step-up voltage regulator: a switch-mode voltage regulator that can convert an input voltage to a higher output voltage ($V_{out} > V_{in}$).
- Simplified step-up regulator diagram: the transistor (switching component) is represented by a switch driven by an error amp.

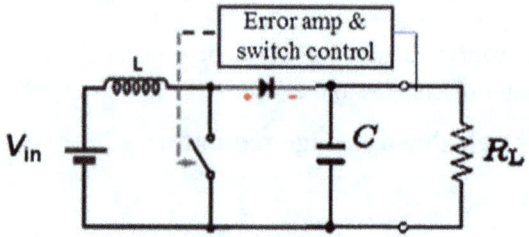

Figure 11.12   Step-up regulator

Step-up voltage regulator – working principle

- When the power switch is on:
  - The diode is RB (diode is off).
  - The inductor stores energy.
  - The capacitor discharges (the longer the switch is on, the lesser the V_{out}).

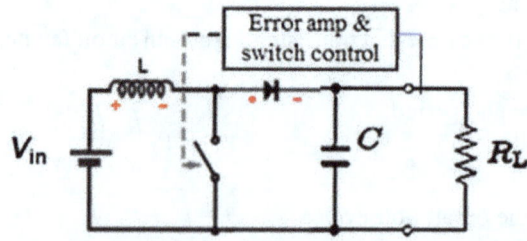

Figure 11.13   The diode is reverse biased

- When the power switch is off:
  - The diode is FB (diode is on).   The inductor changes polarity $\left(V_L = -\dfrac{dL}{dt} \right)$

- – The capacitor charges.
- – The inductor voltage V_L adds to the input voltage to step up the output voltage.

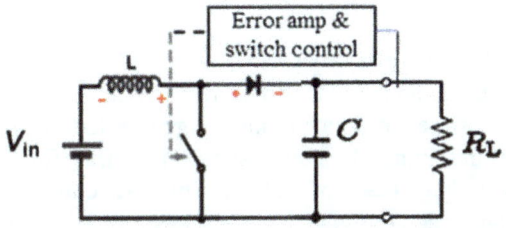

Figure 11.14   The diode is forward biased

11.3.3   Inverting voltage regulator

- Inverting voltage regulator (fly back or boost-buck voltage regulator): a switch-mode voltage regulator that output voltage is of opposite polarity of the input signal (V_{out} and V_{in} out of phase).
- Boost and buck: the inverting regulator can be used to raise (boost) or lower (buck) the positive input voltage to the negative output voltage.
- Simplified inverting voltage regulator diagram: the transistor (switching component) is represented by a switch driven by a pulse width modulator error amp.

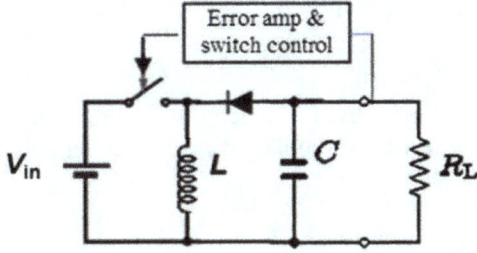

Figure 11.15   Inverting regulator

Inverting voltage regulator – working principle

- When the switch (transistor) is on:
  - – The diode is RB (diode is off).
  - – The input voltage is connected to the inductor, and the inductor stores energy.
  - – The capacitor discharges to the load.
- When the switch (transistor) is off:
  - – The inductor changes polarity.
  - – The diode is FB (diode is on). $$V_L = -\frac{dI_L}{dt}$$
  - – Energy is released from the inductor to the capacitor (charges) and the load producing a negative output voltage.

- Smooth V_{out}: the inductor opposes any change in current to smooth the waveform at the output and keep the output relative constant.

Summary

Op-amp voltage regulators

- Voltage regulator: a device that is designed to automatically maintain a stable output voltage despite variations in input signal or load.
- One method of obtaining DC power supply is to transform, rectify, filter, and regulate an AC voltage (each of that performs a specific function).
- The stages of a power supply:

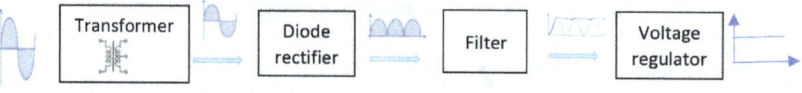

Figure 11.16   DC power supply

Line regulation and load regulation

- Line regulation: $V_{in} \rightarrow V_{out}$

- Load regulation: $R_L \rightarrow V_{out}$

Percent regulation

- Percent regulation can be used to specify the performance of a voltage regulator. It can be in terms of line regulation or load regulation.

- Percent line regulation: $\left(\dfrac{\Delta V_{output}}{\Delta V_{input}} \right) 100\%$

- Percent load regulation: $\left(\dfrac{V_{No\ load} - V_{Full\ load}}{V_{Full\ load}} \right) 100\%$

Classification of voltage regulators

- Linear voltage regulator: a voltage regulator that a linear component (such as a transistor stays in the active/linear region of its operation) during voltage regulation is used to regulate the output voltage.
- Switching voltage regulator: a voltage regulator that a switching component (such as a transistor operates in cut-off or saturation state) during voltage regulation is used to regulate the output voltage.

Table 11.1   Linear voltage regulators

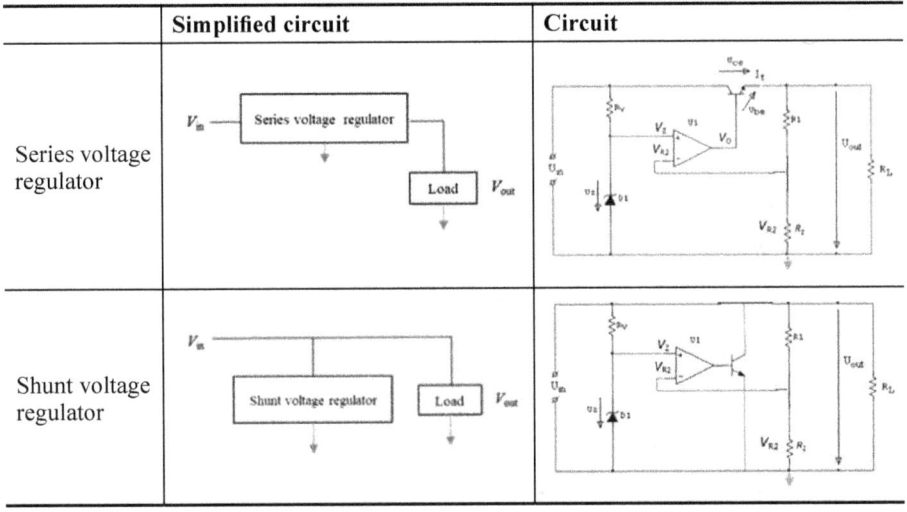

	Simplified circuit	**Circuit**
Series voltage regulator		
Shunt voltage regulator		

Table 11.2   Switching voltage regulators

		Simplified circuit
Step-down voltage regulator	$V_{out} < V_{in}$	
Step-up voltage regulator	$V_{out} > V_{in}$	
Inverting voltage regulator	V_{out} and V_{in} out of phase	

Series voltage regulator

- Output voltage: $V_{out} \approx V_Z \left(1 + \dfrac{R_1}{R_2}\right)$

- Negative feedback in series voltage regulator

$$V_{out}\downarrow \;\to\; V_{R2}\downarrow \to (V_Z - V_{R2})\uparrow \;\to\; V_o\uparrow \;\to\; I_B\uparrow \;\to\; I_C\uparrow$$
$$V_{out}\uparrow \;\longleftarrow\;$$

- Working principle:

$$V_{in}\nearrow \;\; \text{or} \;\; R_{Load}\nearrow \;\;\to\;\; V_{out}\nearrow \;\to\; V_{R2}\nearrow \to\; V_{R2}=V_Z^{\rightarrow}$$
$$V_{out}^{\rightarrow}\;\longleftarrow\;$$

Shunt voltage regulator

- Negative feedback in the shunt op-amp voltage regulator:

$$V_{out}\downarrow \;\to\; V_{R2}\downarrow \to V_o\downarrow \;\to\; I_B\downarrow \;\to\; I_C\downarrow = V_{CE}\uparrow$$
$$V_{out}\uparrow \;\longleftarrow\;$$

- Output voltage and period:
  - Output voltage: $V_{out} = V_{in}\left(\dfrac{t_{on}}{t_{on}+t_{off}}\right)$
  - Period of the on-and-off cycle: $T = t_{on} + t_{off}$

Step-down voltage regulator

- When the power switch (transistor) is on: the diode is reverse biased (RB) (diode is off) $\to C$ charges
- When the power switch (transistor) is off: the diode is forward biased (FB) (diode is on) $\to C$ discharges
- Negative feedback in the step-down voltage regulator:

$$V_{out}\downarrow \;\to\; V_{R2}\downarrow \;\to\; V_o\downarrow \;\to\; t_{on}\uparrow$$
$$V_{out}\uparrow\;\longleftarrow\;$$

Step-up voltage regulator

- When the power switch is on:
  The diode is RB (diode is off) $\to L$ stores energy $\to C$ discharges.
- When the power switch is off:
  The diode is FB (diode is on) $\to C$ discharges $\to V_L$ add to V_{in} to step-up V_{out}

Inverting voltage regulator

- When the switch (transistor) is on:
  The diode is RB (diode is off) $\to L$ stores energy $\to C$ discharges to the load
- When the switch (transistor) is off:
  The diode is FB (diode is on) $\to L$ releases energy $\to C$ charges $(-V_{out})$

Self-test

11.1

1. A voltage regulator is a device that is designed to (   ) maintain a stable output voltage despite variations in input signal or load.
2. One method of obtaining DC power supply is to transform, rectify, (   ), and regulate an AC voltage.
3. A (   ) is a device that converts two-directional AC voltage into single-directional DC.
4. A (   ) is a device that smooths the voltage waveform at the output of the rectifier from varying greatly to a small ripple.
5. The (   ) regulation is the ability of a voltage regulator to maintain the output voltage V_{out} level with a varying load.
6. The (   ) regulation can be used to specify the performance of a voltage regulator. It can be in terms of line regulation or load regulation.
7. A (   ) voltage regulator is a voltage regulator that a switching component during voltage regulation is used to regulate the output voltage.
8. A (   ) voltage regulator is a voltage regulator that is connected in parallel with the load.
9. A (   ) voltage regulator is a switch-mode voltage regulator that can convert an input voltage to a higher output voltage.

11.2

10. A series voltage regulator is a voltage regulator that uses a control element connected in series with the (   ).
11. The resistors R_1 and R_2 in a series voltage regulator help in sensing the variations in the output and provide a (   ) signal to op-amp.
12. The shunt voltage regulator regulates output voltage by shunting (   ) away from the load to regulate the output voltage.

11.3

13. A step-down voltage regulator is a (   )-mode voltage regulator that can convert an input voltage to a lower output voltage.
14. The (   ) waveform is the waveform repeats itself with the repetition on and off action of the switch.
15. For the step-down voltage regulator of Figure 11.17, determine the output voltage, if the input voltage is 15 V, on time is 4μs, and off time is 5μs.

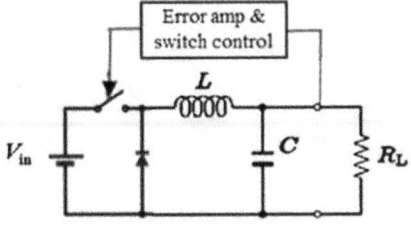

Figure 11.17   Ch 11: No. 15, self-test

16. An (   ) voltage regulator is a switch-mode voltage regulator that output voltage is of opposite polarity of the input signal.

17. The (   ) regulator can be used to raise or lower the positive input voltage to the negative output voltage.

Appendices

Future Trends and Wrap Up

Need of electronic circuits and devices

- The beginning of electronics has left a remarkable impact on human society. Nowadays, we can see that electronic devices are found in various applications such as industries, health care, communication, automotive, consumer, military, aerospace, etc., in which these applications are impossible to function without electronic devices and gadgets.
- Since the increase in interest and the rise of computer technology, artificial intelligence, advanced automation, advanced material, etc., electronic circuits and devices are playing an important fundamental role in the digital age.

The future trends in electronic circuits and devices

- Electronic circuits and devices have made and continue to make incredible contributions to most aspects of human society – a truth that cannot be neglected. Moreover, it may have a bigger impact on human civilization in the future.
- The demand for the latest electronic devices has been trending and will continue to grow. Electronics industries must progress fast to the upswing by using advanced technologies such as:
  – Advanced automation
  – Smart and autonomous systems
  – Quantum computing
  – Nano-electronics
  – Silicon carbide electronics
  – Advanced material
  – Organic electronics
  – Miniaturization
  – Artificial intelligence
  – Robotics
  – Bioelectronics
  – Cognitive science
  – Cloud computing
  – Big data and analytics
  – Computational biology and bioinformatics

– Internet of things (IoT)

IoT refers to the billions of physical objects—"things" around the world that are connected to the internet for the purpose of connecting and exchanging data.

– ... ...

Wrap up

- Electronics has a lot of scope in the times to come, every day as you look around you can find yourselves surrounded by different electronic devices and gadgets. Without them, our daily lives seem impossible.
- Technology is advancing quickly in the field of electronics and these emerging and upgraded electronic technologies and applications will be the growth drivers for the industry, and it will change the way we live and shape the future.

Answers to Self-Test Questions

Chapter 1
1. proton
2. protons
3. 8
4. 18
5. 4
6. semiconductor
7. depletion
8. DC
9. increase
10. forward
11. 0.3
12. increase
13. cathode
14. cathode
15. cathode
16. reverse
17. forward

Chapter 2
1. constant
2. doping
3. $Z_Z = 9\Omega$
4. series
5. $V = 71$ mV
6. voltage shifter
7. clipper
8. regulator
9. both
10. line
11. $V_{in\,(min)} = 4.608$V, $V_{in\,(max)} = 15.108$ V
12. $I_{L(max)} \approx 0.14$ A , $R_{L(mim)} \approx 50\ \Omega$
13. percent voltage regulation ≈ 7.14 %
14. Zener

15. capacitance
16. forward
17. color
18. green
19. dark
20. photo
21. electrical
22. metal
23. Schottky

Chapter 3

1. rectifier
2. regulator
3. transformer
4. center-tapped
5. reverse
6. $V_{pk(out)} = 54.3$ V,   PIV $= 55$ V
7. secondary
8. $V_{pk(sec)} = 240$ V, $V_{pk(out)} = 119.3$ V ,   PIV $= 239.3$ V, $V_{avg} \approx 76$ V
9. $V_{pk(out)} = 238.6$ V,   PIV $= 239.3$ V
10. DC
11. filter
12. product
13. longer
14. lower
15. ripple
16. clipper
17. $V_{pk\ (out)} \approx 8.355$ V
18. biased
19. negative
20. combination
21. clamper
22. $V_{RMS(out)} \approx -13.65$ V
23. DC

Chapter 4

1. current
2. holes
3. base
4. β_{DC}
5. $\alpha_{DC} \approx 0.98$, $\square_{DC} = 59$
6. $\square_{DC} \approx 87.5$, $\alpha_{DC} \approx 0.989$
7. increase
8. $I_B \approx 287\ \mu A$, $I_C = 43.05$ mA, $I_E \approx 43.34$ mA, $V_{CE} \approx 9.4$ V, $V_{CB} = 8.7$ V
9. collector
10. switching

11.   active
12.   closed
13.   saturation
14.   active
15. $I_C(\text{sat}) \approx 3.7$ mA , $I_C = 12.9$ mA, $I_C > I_{C\,(\text{sat})} \rightarrow$ Saturated
16.   zero
17.   internal
18. $R_C = 1875\ \Omega$
19. $V_{\text{in(min)}} = 4.9$ V   The minimum value of input voltage is 4.9 V.

Chapter 5

1.   current
2.   amplified
3.   linear
4.   cutoff
5.   distortion
6. $IC = 21$ mA, $V_{CE} = 8.7$ V
      The Q-point is close to the center of the DC load line.

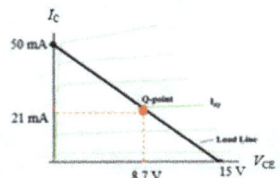

Fig. Ans. Ch 5: # 6

7.   biasing
8.   temperature
9. $I_C \approx 2.87$ mA, $V_{CE} = 0.65$ V
10. $I_C \approx 7.53$ mA, $V_{CE} \approx 14.56$ V
11. $I_{C\,(\text{sat})} \approx 12.5$ mA, $V_{CE(\text{cutoff})} = 30$ V
12. temperature
13. $I_C \approx 1.26$ mA, $V_{CE} = 2.4$ V

Chapter 6

1.   internal
2.   emitter
3.   mho
4.   datasheets
5.   small
6.   base
7. β_{ac}
8. γ_e
9.   open

10. short
11. increases
12. $I_C \approx 1.61$ mA, $V_{CE} \approx 6.78$ V, $R_{in} \approx 1.16$ kΩ, $V_b \approx 5.61$ mV, $R_o = 1$ kΩ

    A_V without C_E: $A_V \approx 0.99$

    A_V with C_E (without R_L): $A_V \approx 64.39$

    A_V with C_E and R_L: $A_V \approx 62.85$
13. 180^0 out of
14. collector
15. input
16. $R_{in} \approx 1.69$ kΩ, $R_o \approx 19.92$ Ω, $A_V \approx 0.98$
17. gain
18. emitter
19. cascode
20. $A_{VT} \approx 8273.6$, A_{VT} (dB) ≈ 78.15 dB

Chapter 7

1. voltage
2. depletion
3. holes
4. gate
5. negative
6. pinch off
7. transfer
8.

V_{GS}	$I_D = I_{DSS} (1 - \dfrac{V_{GS}}{V_{GS(off)}})^2$	Ordered Pair (V_{GS}, I_D)
0	$I_D = (4 \text{ mA})(1 - \dfrac{0}{-7V})^2 = 4$ mA	(0, 4)
-1 V	$I_D = (4 \text{ mA})(1 - \dfrac{-1V}{-7V})^2 \approx 2.94$ mA	(-1, 2.94)
-2 V	$I_D = (4 \text{ mA})(1 - \dfrac{-2V}{-7V})^2 \approx 2.04$ mA	(-2, 2.04)
-4 V	$I_D = (4 \text{ mA})(1 - \dfrac{-4V}{-7V})^2 \approx 0.74$ mA	(-4, 0.74)
-6 V	$I_D = (4 \text{ mA})(1 - \dfrac{-6V}{-7V})^2 \approx 0.08$ mA	(-6, 0.08)

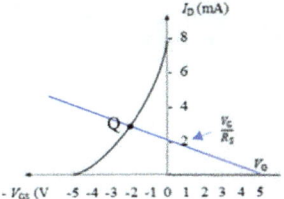

Fig. Ans. Ch 7: # 8

9. drain
10. $V_{GS} = -0.8$ V, $V_{DS} = 3.2$ V
11. Q point: $(V_{GS}, I_D) \approx (-2$ V, 4 mA)

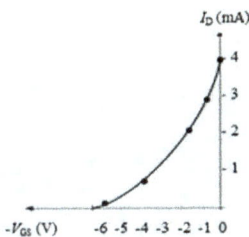

Fig. Ans. Ch 7: #11

12. 0.5
13. Voltage-divider
14. (a) $V_D = 11$ V, $V_S = 10$ V, $V_{GS} = -5$ V, $V_{DS} = 1$ V
    (b) $(V_{GS}, I_D) \approx (-2$ V, 3 mA)

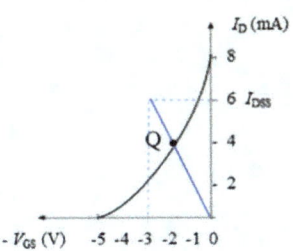

Fig. Ans. Ch 7: #14

15. enhancement
16. Depletion
17. insulating
18. collector
19. $V_{GS} = 6$ V
20. $I_D = 110$ mA, $V_{DS} = 1.8$ V
21. $I_D = I_{DSS} = 6$ mA

Chapter 8

1. gate
2. source
3. drain
4. drain
5. source
6. source
7. input
8. $R_{in} = 9 \text{ M}\Omega$, $R_{out} \approx 0.32 \text{ k}\Omega$, $A_V \approx 0.95$
9. 180^0 out of
10. $R_{in} \approx R_G = 4 \text{ M}\Omega$, $R_{out} \approx 1 \text{ k}\Omega$, $A_V \approx 3.56$
11. in
12. $R_{in} \approx 0.24 \text{ k}\Omega$, $R_{out} \approx 5 \text{ k}\Omega$, $A_V \approx 15.65$

Chapter 9

1. negative
2. zero
3. Negative
4. input signals
5. inverting
6. difference
7. 180^0 out-of
8. $V_{out} \approx -1600 \text{ mV}$, $Z_i = R_1 = 1 \text{ k}\Omega$, $Z_o \approx 8.9 \text{ m}\Omega$
9. $V_{out} \approx 840 \text{ mV}$, $Z_i \approx 6.192 \text{ G}\Omega$, $Z_o \approx 6.46 \text{ m}\Omega$
10. buffer
11. $V_{out} = 30 \text{ mV}$, $Z_i \approx 144 \text{ G}\Omega$, $Z_o = 250 \text{ }\mu\Omega$
12. clipping
13. inputs
14. step
15. zero
16. inverting
17. unity
18. $V_{out} \approx -104.2 \text{ mV}$
19. noise
20. voltage
21. $V_{out} \approx 1.5 \text{ mV}$
22. inverting
23. square
24. $V_{out} = -27 \text{ V}$
25. area
26. triangular
27. When $V_{in} = +8 \text{ V}$: $\Delta V_{out} \approx -12 \text{ V}$
    When $V_{in} = -8 \text{ V}$ $\Delta V_{out} = 12 \text{ V}$

8 V

-8 V

12 V

-12 V

Fig. Ans.(a), Ch 9: #27   *Fig. Ans.(b),   Ch 9: #27*

Chapter 10

1. AC
2. feedback
3. relaxation
4. low
5. tuned
6. positive
7. positive
8. flywheel
9. damped
10. one
11. transformer
12. tickler
13. LC
14. inductor
15. tank
16. capacitor
17. piezoelectric
18. piezoelectric
19. capacitor
20. feedback
21. resonant
22. gain
23. $C = 994$ pF, $R_2 = 50$ kΩ

Chapter 11

1. automatically
2. filter
3. rectifier
4. filter
5. load
6. percent
7. switching
8. shunt
9. boost
10. load
11. feedback
12. current

13. switch
14. repeating
15. $V_{out} \approx 6.67$ V
16. inverting
17. inverting

Index